Programming and Customizing the PIC Microcontroller

Programming and Customizing the PIC Microcontroller

Myke Predko

McGraw-Hill

New York San Francisco Washington, D.C. Auckland Bogotá
Caracas Lisbon London Madrid Mexico City Milan
Montreal New Delhi San Juan Singapore
Sydney Tokyo Toronto

Library of Congress Cataloging-in-Publication Data
Predko, Myke.
   Programming and customizing the PIC microcontroller / Myke
Predko.
    p.   cm.
  Includes index.
  ISBN 0-07-913645-1 (hc).—ISBN 0-07-913646-X (pbk.)
  1. Programmable controllers. I. Title.
TJ223.P76P74 1998
629.8'9—dc21   97-21762
                                        CIP

McGraw-Hill

A Division of The **McGraw·Hill** *Companies*

  2 3 4 5 6 7 8 9 0   DOC/DOC   9 0 2 1 0 9 8

P/N 052582-X (HC)
PART OF
ISBN 0-07-913645-1
and
P/N 052583-8 (P)
PART OF
ISBN 0-07-913646-X

The sponsoring editor for this book was Scott Grillo, the editing supervisor was
John Baker, and the production supervisor was Sherri Souffrance. It was set in ITC-
Century Light by Wanda Ditch through the services of Barry E. Brown (Broker—
Editing, Design and Production).

Printed and bound by R.R. Donnelley & Sons Company.

McGraw-Hill books are available at special quantity discounts to use as premiums
and sales promotions, or for use in corporate training programs. For more
information, please write to the Director of Special Sales, McGraw-Hill, 11 West
19th Street, New York, NY 10011. Or contact your local bookstore.

Contents

8 Experiments *165*

9 Projects *237*

Acknowledgments

Just as no book is ever written in one day, no book is written by a person in a vacuum. There are a number of people and companies that I would like to acknowledge, without whose help, this book would not be written.

The first "Thank you" goes to everyone on the PIC and STAMP Lists over the past few years. Demonstrating what a powerful force the Internet can be for goodness, the 2000 or so individuals who each monitor the flurry of mail that comes through each day and post replies, suggestions, ideas, and the occasional joke have made learning the PIC and developing applications for it a real joy. While I could fill several pages with the names of individuals who have helped in providing information used in this book and other projects that I have done, I'm going to refrain, in fear of missing somebody.

Normally, with a technical book on a chip or processor, such as this one, the device's manufacturer goes out of its way to make sure the author has all relevant information and samples required for the book. I did not put this request to Microchip, as their normal technical support has been terrific and responses to questions were received usually within one day. Throughout this project, Microchip has demonstrated superior customer support, both to the small and large customer as their SOP. I also would like to thank Microchip for their generosity in allowing the inclusion of the MPASM and MPSIM programs along with the numerous illustrations used in this book (which meant I didn't have to recreate them).

Like Microchip, there are two companies that I would like to recognize and thank for making superior service something their customers can depend upon. Digi-Key and AP Circuits made the job of developing the projects used in this book a lot easier by always delivering on time with exactly what was ordered. The Programmer and Emulator (and really all the other projects) given in this book would not have been possible without these companies' reliable service and help in getting datasheets for parts or explaining exactly what an "Aperture" file was.

I would like to also thank a number of individuals without whose help this book would not be as good as it is.

At the top of the list is Ben Wirz, for discussing the various projects and their requirements with me, as well as making sure that they are manufacturable. Sorry for bending your ear and sending all those long e-mails. One day, we'll get together for that beer.

Along with Ben, I would like to thank Don McKenzie, Kalle Pihlajasaari, and Mick Gulovsen (a.k.a. the WE Dealers) whose suggestions and support were instrumental in developing the requirements for the Emulator and other projects we have worked on together. A lot of the stuff here wouldn't exist without your help.

I would like to thank everyone at Celestica (my regular employer) for helping to make this book happen. A lot of people were called upon to answer questions about interfacing to different devices, PCs, and other equipment. Karim Osman should be singled out for his help in making sure the programmer presented in the book would work on a very wide variety of devices.

I appreciated the efforts of Dave Cochran of Pipe-Thompson (my local Microchip Rep) who made sure that I always had everything I needed and all my questions were answered.

Thanks to my children—Joel, Elliot, and Marya—for their excitement and enthusiasm for this project.

And finally, the biggest "Thank You", has to go to my wife, Patience. Thank you for letting me spend all those hours in front of the PC, listening to me puzzle out strange sounding things like "eye-squared-see," making suggestions on the projects, tolerating the weekly visits by FedEx, Purolator, and UPS, and not running away when I asked for some quick proofreading. Writing something like this book is an unbelievably arduous task and never would have been possible without your love and support.

Let's go lie out on a beach.

myke predko
Toronto, Canada
February, 1997

Introduction

As you grow older, most things that you remember will become cloudier and less clear ("I dated who?"), while others will stay as vividly in your memory as though they just happened. These snippets of what happens around us are vitally important in shaping us as human beings. They are known as "Significant Emotional Events" and, as a psychology professor I once had described, they are largely responsible for defining who we are.

I always felt that I received a SEE (as a Significant Emotional Event is known to psych profs) when I first opened up a copy of the IBM Personal Computer Technical Reference and saw the schematic for the keyboard section of the PC. The keyboard was simply a switch matrix wired to a single box, labeled "I8048," which was in turn wired to the PC itself. Finding a copy of Intel's "Microcontroller Data Book," I discovered the 8048 was a chip with a microprocessor, memory, interrupt control, and input/output—an analogous computer system that was communicating to the main part of the PC, which took up 20 or so pages of schematics!

Reading through the databook, I marveled at how much the little creatures could do. My mind was abuzz with potential applications, and I started listing them out on paper.

That was a long time ago, and still my philosophical wonderment for microcontrollers hasn't died down. In the years since first looking at that schematic, I've had an opportunity to design a few applications using microcontrollers and learn what their capabilities are.

So what is a microcontroller? If you remember your first instruction in computer science, you'll remember the diagram shown in Fig. I-1.

A microcontroller contains these three basic functions (along with a few extra) all contained on one chip.

Microcontroller applications

Where can you find a microcontroller? Well, just about anywhere you look. If you could pick up your car and shake it, I'm sure a number would fall out. In the car, they

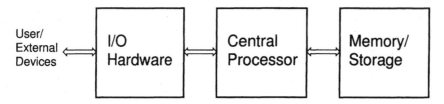

I-1 Basic computer block diagram.

are used for controlling the engine, the brakes, the radio, and even the seats. Your kitchen appliances have so many that the joke of the talking toaster isn't all that far-fetched. If you remember back about 15 years to the Air Canada Boeing 767 that glided down to a drag racing strip after running out of fuel, the aircraft had 69 microcontrollers that all stopped working when the engine-driven generators stopped.

After this list, you're probably thinking that something used in a toaster would be radically different from what is used in an airliner. Actually, they aren't.

While there is a myriad of different devices out there, they can all be modelled using the basic diagram in Fig. I-1. Learning about one chip, will give you understanding as to how all microcontrollers are designed into applications.

The decision to determine which microcontroller to use is answered by making a list of requirements that you have for the application. Before choosing which type of microcontroller to use, a list of requirements should be made. This list would then be matched to various microcontroller descriptions, and the one that was best suited could be identified easily. However, before you are able to do this, you should really have an idea of what microcontrollers can do.

This turns into a chicken and egg question when you ask which microcontroller should you start with? If you go out and search manufacturer's Data Books, you will be bombarded with different types of devices with a myriad of different features. When you compare what each manufacturer offers, you'll find that, when the total range of the devices is examined, you can pretty much find similar features across the product lines. The decision on what device to use will really boil down to such points as cost, availability, support, and what you are familiar with.

I've picked a family of devices that contains features that are fairly common to devices in general and has some advantages to allow you to experiment with it easily and without breaking your budget. The tools necessary for doing all the applications in this book should be sitting on your workbench right now.

The microcontroller family discussed in this book is the "PIC" Series by Microchip Inc. (Microchip's address can be found in appendix D.) The devices contain a RISC (Reduced Instruction Set Computer) Processor, anywhere from 256 bytes to 16K of Instructions of Program Space, tens to hundreds of bytes of Variable RAM, and some number of I/O bits (which can have additional advanced features). Like most microcontrollers, the features that make this device so attractive for use in a very wide range of applications include it's robust design with regards to Power, Clocking, and Reset.

For this book, I've primarily focused on the PIC16F84, which has Electrically Erasable PROgrammable Memory ("EEPROM") for program storage. This allows simple changes to the contents of memory without having to invest in ultra-violet erasers. The software development tools used in this book are available free of charge from Microchip.

Actually programming the devices can be accomplished either by various manufacturers' programmers, or by using the Programmer circuit that is included in this book. The programmer in this book has many of the features of other low-cost programmers that are available, but with the important difference that it is capable of supporting a wide variety of different hosts (and not just the IBM PC). I have included a number of sources for programmers in appendix D.

Also included in this book is the design for a simple PIC emulator. As I have designed the emulator, I can really see how useful it can be for faster development and debugging of applications. Like the programmer, the emulator has a simple RS-232 Interface that allows its use with a wide variety of hosts. While it is very low-cost, there are two major limitations on its usefulness in helping to debug applications: The emulator is relatively slow and is unable to use symbolic information from the source code. However, even with these limitations, it is an excellent tool to help understand how the PIC works.

As you go through this book, I've tried to build knowledge in a "bottoms up" format, in the following order:

- The features of the microcontroller
- Application design notes and rules
- Investigating the microcontroller
- Sample applications and projects
- Turning the reader loose

Please don't think that I've written a book on just the PIC microcontrollers (and the 16F84 specifically). I've tried hard to provide a framework that anybody can use to investigate a microcontroller or choose the appropriate one for a given application. I hope you don't feel I've given your favorite microcontroller the short shrift, because that's not my intention at all with this book. What I've tried to do is pass along some of the knowledge I've picked up over the years along with a few of the fun things I've created. I've tried to keep this book (well) below the level of a graduate engineer.

As this book went to press, the original EEPROM based PIC (the PIC16C84) became "obsolete". This was a mainstay device for learning about the PIC and developing first applications on. The 16C84 has been replaced by the 16F84, which has more file registers and some differences in the configuration fuses, by Microchip and will be the primary device used in this book.

This book is not meant to explain electrical engineering and computer science basics. Concepts and ideas will probably be touched upon in this book that aren't fully explained or obvious. If you do discover something that you are not sure of, you can find full explanations in your local library.

I hope you find this book useful, both in terms of learning about the PIC, as well as giving yourself an idea of what microcontrollers can do in general.

PIC resources

In this book, you will see numerous references regarding where to get software and information. The latest revisions of this data can be found on the Internet. I have tried to ensure that only resources that are on the 'Net are used in this book because of the Internet's common and consistent manner of accessing data.

All of the tools you will find listed in the appendices are for PC-DOS machines. Microcontrollers generally do not require a powerful development system, and the PC is well documented and easy to use. There are Internet sources with source code for the various tools that you might be interested in porting to other platforms.

While there are numerous sources for assemblers, compilers, simulators, and programmers, I have selected and included Microchip's PC-DOS command-line development tools for this book ("MPASM" and "MPSIM") along with the programmer and emulator. This means that the book contains all of the tools required to develop (18-pin) PIC applications.

At the end of this book, I have listed a number of sources for information and products. I consider the "list groups" to be the most important resource you can have. "List groups" are informal programs that distribute Internet e-mail to a list of people. This e-mail usually consists of questions and answers, information, and product updates. I have found this to be truly the best single resource that you can have for PIC information.

Conventions used in this book

KΩ—1000 ohms

μF—Microfarads

ms—Milliseconds

μs—Microseconds

0x0*nnn*—Hex number

0b0*nnn*—Binary number

nnn—Decimal number

AND, OR, XOR—Bitwise logical operations

_*Label*—Negative active signal/bit

Register.Bit—Single bit in a register

`Monospace font`—Code examples

1
CHAPTER

Microcontrollers

One of the most difficult decisions you will have to make when developing an application that requires a microcontroller is deciding which microcontroller to use. In addition to the plethora of different manufacturers out there, each manufacturer has a number of different devices, each designed with different features making them suitable for different applications.

Different types of microcontrollers

Pretty well every manufacturer of microprocessors has a line of microcontrollers. These devices might or might not use the same architecture as their line of microprocessors. To make deciding which part to use harder, each type of device has a number of different part numbers, each with slightly different features.

These features include:
- Programmable Digital Parallel Input/Output
- Programmable Analog Input/Output
- Serial Input/Output (i.e., Asynchronous, Synchronous, and Device Control)
- Pulsed Waveform Output for Motor/Servo Control
- Interrupt on Input conditions
- Timers and Interrupt on Timers
- External Memory Device Interfaces
- External Bus Interfaces (i.e., PC ISA)
- Internal Program Storage Type Options (ROM, EPROM, PROM, and EEPROM)
- Internal RAM Options
- Floating Point Computation

And the list goes on and on.

Further complicating the list is the variety of different devices that might be considered to be microcontrollers. These include single chip PCs and Digital Signal Processors (DSPs).

This means that you have an almost overwhelming choice in what types of microcontroller you can use for your application. This book is about the PIC and what you can do with it.

The PIC microcontrollers

As noted in the introduction, a requirements list is necessary when it is time to determine which microcontroller is appropriate for a given application. When choosing a microcontroller for an application, you should make sure that the device you choose meets all your requirements including cost. You might want to substitute a device with fewer hardware items if a cheaper version is available and the hardware feature can be emulated in software.

When I was presented with this problem, I looked around for the product line of microcontrollers to focus on. I looked for the following parameters:

- Simple Power, Reset, and Clocking requirements
- Inexpensive software development tools
- Large user base
- Readily available parts
- Ease of programming using hobbyist equipment

All of these parameters were important because I didn't want to get into something that would cost a fortune just investigating and because I wanted to avoid specifying parts that would be hard to procure.

What actually led me to the Microchip PIC was seeing articles on the Parallax Stamp that were coming out in hobbyist magazines. I'm a real "bottoms up" type of engineer, and I wanted to intimately understand the chip that was driving the Stamp.

As I looked into the PIC, I was impressed by the wide variety of features available in the different devices, the large user base already established, and the amount of support from Microchip, ranging from emulators and programmers to free software development tools and a large selection of application notes.

The PICs themselves use a common core. This means there is a large amount of commonality between the devices in terms of software and hardware interfacing. In looking through the list, you can see that there is a lot to consider when choosing a device.

The PIC Processor core is built around the "Harvard architecture." Once upon a time, the U.S. Department of Defense asked Harvard and Princeton to come up with a design for a computer for plotting artillery shell trajectories. Harvard came up with an architecture that used two memories: one for storing the program (I call it "Control Store" and will often use this term in the book) and the other for variables ("RAM"). The advantage of the design was that instructions could be fetched from (the primitive and slow) Control Store at the same time that RAM (also primitive and slow) was accessed, speeding up instruction execution time. The competitive architecture (from Princeton, known as the "Von Neumann architecture") used the same memory for Control and RAM storage. This concept of breaking up the variable RAM and Control Store made the Harvard architecture much faster than the Von Neumann, but it was more expensive as well. The Princeton architecture won out and eventually became the central architecture for most computers (i.e., the PC on your desk).

Many years after the competition, some designers at General Instruments, who wanted to create a microcontroller design of their own, resurrected the Harvard architecture. Having a separate Control and RAM Store has some significant advantages for microcontrollers, meaning that the design of the device can be made simpler and considerably faster (both important considerations for microcontrollers). Actually, Harvard-architecture devices are a bit cheaper to make than standard Von Neumann devices because it is not necessary to mix ROM (EPROM, etc.) with program RAM. Using the Harvard architecture makes the PIC line of microcontrollers considerably faster and cheaper than their competition.

That was in the mid-1970s. Now Microchip (the owner of the General Instruments line of products) has a very complete line of CMOS PIC microcontrollers with a plethora of features that make the devices suitable for a wide range of applications.

In terms of the different types of PICs, I like the way Microchip has broken the devices into three groups: the "low end," "mid-range," and "high end." All of the devices are upward code compatible (i.e., the higher-range devices can run code designed for the lower-range devices) and have registers located at the same address locations to simplify moving code between devices.

The examples used in this book use the PIC16F84, while the projects and applications presented in the book use a fair range of PICs to show how different features within the PICs are used. The 16F84 itself is pretty low-end when its features are compared against what some of the other PICs have. The reason that I choose this device is because of the EEPROM Control Store memory (which can be rewritten without requiring an ultra-violet EPROM eraser) and its use of interrupts, which increases the devices capabilities significantly.

The Parallax Basic Stamp

One device that should be mentioned is the Parallax Basic Stamp. The Basic Stamp is a PIC16C56 with a 2K Bit Serial EEPROM attached to it. It is really a marvelous device, being programmable (and accessible) via RS-232. The Stamp is programmed in PBASIC, which is a high-level language designed to reduce the complexity of application development. There are numerous applications designed for the Stamp.

The name of the "Stamp" comes from the fact that the actual hardware is about the size of a postage stamp.

The Stamp's source code is compiled from PBasic into tokens. These tokens are downloaded into the Stamp and are stored in the EEPROM. Each line is converted into a token of two or three bytes. This token is then executed by the PIC. With the token system, the Basic Stamp can store 60 to 70 lines of code.

The Stamps have 8 bits of digital I/O. However, through functions in the PBASIC language, Basic Stamps can carry out a number of advanced operations, including serial I/O and analog measurements. The Stamp serial I/O function can be used to allow the application code to communicate with a TTY device or allow a controlling PC to monitor and debug the code and hardware application, providing an interactive environment for the developer.

There is an update to the Basic Stamp from Parallax called the Basic Stamp II (or BSII). This device offers increased program storage, faster program execution, more built-in instructions, and more I/O pins. In addition to the BSII, there are a number of companies that make and sell "clone" Stamps. These copies are often sold with faster clocks and larger EEPROM memories than the original Basic Stamp for increased application performance.

The Basic Stamp has much poorer performance than what would be expected of a stand-alone PIC programmed with the application. To rectify this, there are a number of PBASIC compilers beginning to become available on the market. Compiling the source code means that standard PICs can be used in the place of the Basic Stamps. The compiled code runs considerably faster on the PIC than on the Basic Stamp (usually on the order of 10 times faster). The languages allow a Basic Stamp to be used as the application debug tool, giving the developer the advanced debugging functions of the Stamp and, when the application is ready for release, the cheaper, faster PIC can be put in its place.

The reason that I've avoided the Stamp in this book is because a lot of the PIC hardware is hidden by the Stamp Programming language. However, the Stamp is an excellent tool to introduce yourself to microcontrollers. Sources for Basic Stamp information are listed in Appendix D.

2
CHAPTER

PIC hardware features

With the PIC microcontrollers, it is easy to get totally engrossed in the programming of the devices and not understand any of the hardware considerations. Like many topics introduced and discussed in this book, the various aspects of the hardware could be expounded upon for several books of this size.

Different types of PICs

As time goes on, more and more PICs are added to the product line. While I have not investigated each device in the PIC family in this book, there is enough information here for you to look at a list of PIC features, decide which is best for the planned application, then use the selected device.

8-pin PICs

Microchip has recently come out with a number of 8-pin devices for *very* simple applications. These devices provide up to 6 I/O pins along with a variety of advanced features (including ADC).

The PIC12C5xx parts are essentially low-end devices with in-system programming capability. Also included is an internal RC oscillator (meaning no external pins are required to provide a clock to the PIC) and internal reset circuitry. These features allow the PIC to run with Vdd and Ground and up to 6 I/O pins.

The PIC12C67x parts use the mid-range CPU architecture with built-in ADCs.

Low-end PICs

The low-end PICs were first available in the early 1980s. They consist of a basic processor core with no interrupt capability and a very small program counter stack. They do not have any of the advanced I/O features of the mid-range PICs (i.e., built in ADCs, serial ports, Microprocessor Bus Interfaces, etc.). Their big advantage is that they are very cheap (although this advantage is becoming less and less significant as cheaper mid-range devices become available).

Because of the lack of interrupts, the small program counter stack, and a maximum program size of 2K, I tend to shy away from using these devices.

Mid-range PICs

There is an awful lot of variety in the mid-range of the PIC product line. The devices can have anywhere from 13 to 54 I/O ports, multiple timers, PWM I/O, asynchronous serial communication, direct-control LCDs, EEPROM program store, and the list goes on and on. One of the more interesting devices that some PICs are designed for interfacing directly to LCDs.

The basic mid-range CPU consists of 35 instructions, with interrupts and an eight location deep program counter stack. The interrupt sources can be input pin states, timer overflows, serial data input, A/D completion, along with others. A maximum program size of 8K allows very complex applications.

High-end PICs

The 17Cxx devices are enhancements to the basic architecture used in the low-end and mid-range PICs. Able to access up to 64K 16-bit words of memory, they can provide complete system solutions rather than just traditional intelligent hardware control (as the other PICs typically do).

The CPU core allows for much more flexible internal data transfer and provides for multiple interrupt vectors for different interrupt handlers. These features allow much more efficient coding of applications.

Device packaging

The PICs come in a wide variety of packages. In this book, the majority of experiments and projects use the PIC16F84 in an 18-pin Dual In-Line ("DIP") Package. This is a standard package that is 0.300" wide with 9 pins to a side and 0.100" between pins. See Fig. 2-1.

The reason that all the examples and applications in this book use this package is because of simplicity of insertion/removal as well as tools and support products for this type of package.

There are a number of different packages available for the PIC. The first is the ceramic package with a glass window (Fig. 2-2).

This is used with devices that have EPROM control store that can be erased using ultra-violet light. The disadvantage of this package is its cost and the chance that the EPROM can be accidentally erased. After a device is programmed, the window should always be covered with something like an ultra-violet opaque sticker (an old metalized diskette write-protect sticker works well) to identify that it has been programmed and to help prevent unwanted erasures.

The other type of DIP package that is available for the EPROM part is the plastic package similar to the 16C84 DIP package shown in Fig. 2-1. This package is considerably cheaper than the windowed ceramic. The package is appropriate for ROM-based applications (where the control store program is built in at the Microchip factory), for EEPROM-based devices (i.e., the PIC16C84 and PIC16F84) in

2-1 16C84 in DIP plastic package.

2-2 16C73/JW with a glass window.

which light is not required to erase the device, and for EPROM-based devices to be used in mass production, where the program is set and will not be changed. EPROM devices in plastic packages without windows are known as One-Time Programmable (OTP) devices. Before using devices in OTP packages, the code that is to be loaded into the PICs must be fully qualified and debugged.

Windowed devices are identified as "JW" parts. There is one thing that is very important quirk to note with "JW" versus "OTP" PIC operation. If the "JW" part's window is not covered with a completely opaque label (i.e., the metalized write-protect sticker), you might find that, when you go to nonwindowed parts ("OTP"), your application works differently. This is because the light entering the window changes the electrical characteristics of the silicon.

Along with the standard DIP package, PICs are available in a variety of Surface Mount Technology (SMT) packages (Fig. 2-3).

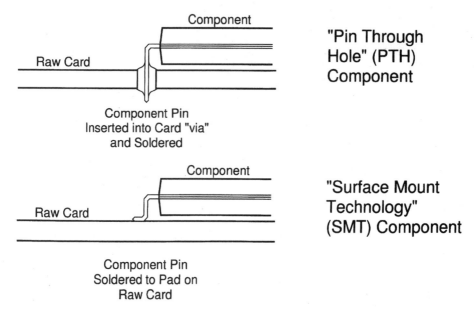

Component

"Pin Through Hole" (PTH) Component

Raw Card

Component Pin Inserted into Card "via" and Soldered

Component

"Surface Mount Technology" (SMT) Component

Raw Card

Component Pin Soldered to Pad on Raw Card

2-3 PTH and SMT solder joints.

Typically, SMT packages are all plastic, which means the devices are OTP by default. The advantages of the SMT packages over the DIP package should be obvious. SMT has a much smaller "footprint" than the DIP device and is shorter in height to allow smaller physical packages for the application. The SMT package also allows components to be mounted on both sides of the board, for increased density. SMT solder joints tend to be of better quality (and easier to inspect) than Pin Through Hole (PTH) joints, resulting in a more reliable and better quality product. Because fewer holes ("vias") are required in the raw card, the SMT package can result in cost savings in the finished product.

After saying all these wonderful things about the SMT packages, you're probably wondering why I don't use these packages for the examples in the book. The simple reason is that they are extremely difficult for the hobbyist to work with. I realize that there are a number of products available for home experimenters to develop applications using SMT, but I feel that, without the resources of a well equipped lab, SMT isn't worth the trouble.

Does this mean I don't think you should ever use SMT? Of course not. Ironically, I'm probably an SMT bigot and feel that these packages should be used instead of PTH packaging. However, I don't feel that SMT is appropriate for learning about these devices. Actually, I don't feel SMT is appropriate for application development because of the difficulty in removing and handling parts, as well as wiring modifications to the board. I feel that SMT is excellent for completed applications that are going into mass production.

You might have an application where you require even smaller devices than SMT. There is one more package type that you might want to consider and that has no packaging at all. Just the PIC chips can be attached and wired directly on the board. This is known as Chip On Board (COB) technology. There are two primary ways of attaching chips directly to the board.

The first is soldering the chip directly to the raw card and running very thin aluminum wires from the pads on the chip to the raw card. This is very analogous to how the chip is put in a standard package. The reason (and advantage) that the chip is soldered to the raw card is that the raw card can be used as a heat sink for the chip.

The ultimate form of COB is the "flip-chip" or "C4" bond attachment process. In this process, the chip pads are soldered directly to SMT pads on the raw card.

Both methods of Chip On Board attachment, while interesting to see and understand, are not possible for the hobbyist, are very expensive for prototype production and are meant for high-volume production in applications where only a very small space is available for the PIC. COB, like other OTP packages, will not allow ultra-violet erasure of the EPROM control store.

The package that you decide upon will have a major impact on the finished application. However, when prototyping, the device that is the easiest to remove and reprogram should be the one that is used.

Control store types

There are three different types of memory used for the program memory (or "control store") of the PIC. They are:

- Erasable PROgrammable Memory (EPROM)
- Electrically Erasable PROgrammable Memory (EEPROM, also known as "Double-EPROM" or "E-Squared PROM")
- Hardwired or Read-Only Memory (ROM)

Each different type of memory offers advantages in different situations.

EPROM is the basic type of control store memory used in the PIC devices.

In an EPROM memory cell, a charge is trapped in an insulator to record the state of the bit. This charge can be removed by applying ultra-violet light to the cell. As noted earlier, PICs with EPROM come in two basic types of packages:

- A ceramic windowed package that can be erased
- Different plastic packages that do not allow the device inside to be erased (OTP)

Mid-range and high-end EPROM (and EEPROM) PICs have the significant advantage of being able to be programmed with the PIC in the application after board assembly. This can be done at either In-Circuit Test (ICT) Functional Test or at a dedicated programming station. The reason why you would want to program the parts after assembly is that it reduces the number of different PICs the card assembler has to keep in inventory. It also can simplify the manufacturing process by eliminating the programming step along with the procedures for handling and identifying the programmed part. This is known as In-System Programming (ISP) and is described at the end of this chapter.

Electrically Erasable PROgrammable memory behaves in a similar manner to EPROM but offers a big advantage. The trapped charge can be electrically cleared (not requiring UV light for erasing), which makes for fast and easy software development and reprogramming. The EEPROM control store can be reprogrammed several hundred times before a failure is expected.

It should be noted that EEPROM memory tends to have slower access times than EPROM or ROM. This might cause problems with high-speed applications.

Currently, the only PIC devices that have EEPROM control store are the $16x8x$ family, which offers functionality that is at a level between the low-end and the higher-featured cards. Right now, there is some confusion in the industry over what is an "EEPROM" part and what is a "Flash" part. Microchip identifies it's EEPROM parts with both a "C" (for "CMOS") and an "F" (for "Flash"), which can be confusing. As well, new EEPROM devices are becoming available. (Nine EEPROM control store PICs are to be released in 1997). The microchip convention for identifying these parts is to use an "F" rather than a "C" in the device's part number.

ROM-based control store means that the program has been built into the chip at the factory. ROM chips cost less than their EPROM and EEPROM counterparts, but there is some overhead that detracts from their attractiveness. The first detractor is the nonrecurring expense (NRE) of getting a ROM mask designed for the product. Without sufficient quantity (10,000 and up), this cost might not be recovered. Having ROM parts built takes time and will cut into your lead time and application debugging. Finding out that there is a problem in the program that you are having built into the part can be very costly to fix. Microchip identifies ROM Control Store with a "CR" instead of a "C" or "F" in the device part number.

For these reasons, I have seen a real trend away from ROM-based applications (both microcontroller and other) and a real trend toward EEPROM/Flash because of the ease of programming at the card assembler (using In-System Programming features) and electrical erasure if a problem with the code is discovered.

When choosing the type of control store for your application, there are the issues of cost, schedule lead time, application speed, and code stability to consider.

Reset

The PIC microcontrollers are reset by a single pin, "_MCLR." When this pin is asserted (i.e., equal to a "0"), the PIC will be in a reset state. This means that all output drivers will be turned off and the local clock (driven by the PIC) is turned off.

The reset vector is the address that the PIC begins execution at the start of the program. This value is 0 for the mid- and high-range PICs and the last address for the low-range PICs. For example, 0x01FF is the reset address of the 16C54.

Reset is not to be confused with power up. On power up, most of the hardware register bits are set to specific values and the remaining bits and the RAM registers have unknown values. After a reset cycle, the remaining bits and the RAM registers are set to the values they were at *before* reset was active. By setting some RAM registers to a known, not easily repeatable state, a program can check to see whether or not the reset vector was executed by a power up or reset condition. In newer PICs, the "PCON" register can be used instead of registers to determine the reason for the reset.

You might want to reset a running PIC if the input voltage goes below a predetermined threshold. This is known as a "brown out." Later in this book, I will show some examples of reset circuitry that will put the PIC into a reset state when the input voltage goes too low.

Normally in the PIC, separate reset circuitry is not required. This is because of the length of time the PIC can be programmed to power up (using the "PWRTE" bit in the configuration fuses).

System clock/oscillators

The PIC can run with a wide variety of different system clocks. This adds a great deal of flexibility to your application's design. In this book, I will use the term *clock* and *oscillator* interchangeably when I'm referring to the clocks used to drive the PICs.

There are three types of oscillators used with the PIC:

- Resistor/capacitor (also known as "RC")
- Crystal
- External

Each has advantages that can be exploited during application development.

Resistor/capacitor (RC) oscillators (Fig. 2-4) use the time constant of a simple R/C network to provide a clock of an approximate frequency (typically 20% accurate).

Note that the OSC2 pin outputs one quarter of the clock frequency (which is actually the instruction clock). This signal can be used to synchronize external devices to the PIC.

This 20% accuracy means that the RC clock circuit can only be used for applications where precise timing is not required. This leaves out applications where the PIC outputs serial data, audible tones, or interfaces with other parts. There still are

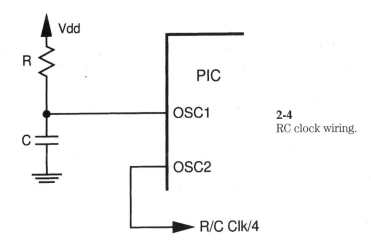

2-4
RC clock wiring.

a lot of applications where the RC clock is appropriate, but the RC clock should not be used in any applications requiring precise timing.

Before using an R/C oscillator, the PIC's configuration fuses must be set appropriately. Other clocking schemes include a low-power clocking mode (LP), standard megahertz-range crystals (XT), and high-speed clocks (HS). These three modes can accept either a crystal (or ceramic resonator) or an external clock source.

The crystal clock circuit typically looks like Fig. 2-5.

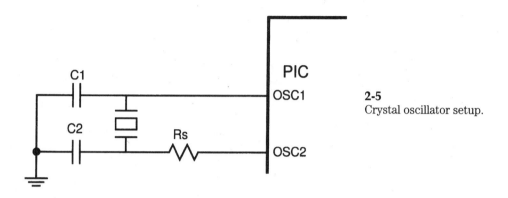

2-5
Crystal oscillator setup.

The external capacitor values (C1 and C2) can be found in the datasheets for the crystal and PIC that you are using, as well as specific to the frequency. The three parameters affect the values for C1 and C2. Rs is a bit of a nebulous value and its inclusion is usually by empirically determined need and is in the hundreds of ohms.

As I was developing applications for this book, I discovered the Ceramic Resonator (and examples of how I use it are throughout chapter 9). This device contains a resonator along with integral capacitors that make the layout of your application simpler (with fewer parts) and potentially cheaper because the Ceramic Resonator typically costs the same as a crystal but does not require the extra parts (capacitors) and vias of a crystal solution. Another bonus is that a Ceramic Resonator is a lot more physically robust than a crystal. The only down side is in terms of timing resolution, typically a ceramic resonator is accurate within 0.5%, where as a crystal with only 20 ppm (0.002%) accuracy is readily available. For most applications, this is not an issue, but for developing real-time clocks and such, a crystal should be used.

To determine how the configuration fuses for the oscillator should be set, you can use the rules shown in Table 2-1, as a rule of thumb.

The system clock starts up at power-on time as shown in Fig. 2-6.

If an external oscillator is used to provide the system clock to the PIC, then the LP, XT, or HS clock mode should be selected. The output of the external oscillator should be put into OSC1, with OSC2 being left unconnected ("floating"). OSC2 will output a signal at the same frequency as OSC1, but it is not within phase of the OSC1 signal. If the clock sent to OSC1 is required for other parts of a circuit, then it should be buffered and fanned out (with OSC1) to the other devices.

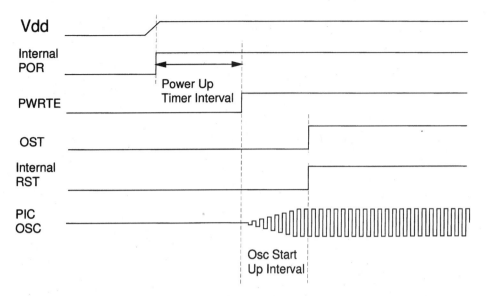

2-6 PIC power-up sequence.

**Table 2-1. Determining how
the configuration fuses for
the oscillator should be set**

Type	Frequency range
RC	0 to 4 MHz
LP	5 KHz to 200 KHz
XT	100 KHz to 4 MHz
HS (−04)	4 MHz
HS (−10)	4 MHz to 10 MHz
HS (−20)	4 MHz to 20 MHz

Hardware and file registers

As explained elsewhere, the PIC has completely separate register and control store spaces. The register space is used for storing values used by the software (also known as "variables") and for locating the PIC hardware control registers. Along with the RAM and registers, there is a "w" register that is used as an accumulator or temporary storage.

The registers are contained within one, two or four 128 or 256 (in the case of the high-range PICs) byte address bank. Each register contains up to 8 bits and can be accessed via a number of different methods in software. Typically, registers that control the hardware and are typically only accessed during the initialization of the application are located in Bank 1. This is an important note because it allows you to structure your initialization to set the bank select and reset it for execution of the program only once (which will save in instructions that are used). See Fig. 2-7.

	Bank 0	Bank 1		
0x000	INDF	INDF	0x080	Notes:
0x001	TMR0	OPTION	0x081	▪ For Higher Function
0x002	PCL	PCL	0x082	PICs, the File
0x003	STATUS	STATUS	0x083	Registers may start
0x004	FSR	FSR	0x084	at 0x020
0x005	PORTA	TRISA	0x085	▪ The Darkly Shaded
0x006	PORTB	TRISB	0x086	Area can be used for
0x007			0x087	either additional Port
				Registers or Special
				Function Registers
0x00A	PCLATH	PCLATH	0x08A	▪ The Lightly Shaded
0x00B	INTCON	INTCON	0x08B	Area Denotes
0x00C			0x08C	unused Regsiters
	File Registers	File Registers		and will return 0x000
				when Read
				▪ The File Registers in
				Bank 1 may be
				Shadowed from
				Bank 0
0x07F			0x0FF	

2-7 Mid-range register map.

The PIC hardware registers are all located at the same addresses. This allows code to be reused easily between PIC devices. Microchip has made efforts in the last few years to make sure that all registers and bits are named in the same manner to help facilitate this. The PIC processor registers (STATUS, INTCON, FSR, INDF, OPTION, PCL, and PCLATH) are all located at the same addresses within each register bank in the PIC. This allows the same access of the different registers, regardless of how the bank select bits are set. The hardware registers start at address zero within each bank and the RAM registers are located immediately following them.

The low-end PICs only have one register bank. To access the registers that are typically located in Bank 1 of the PIC (most notably the TRIS and OPTION registers), special instructions are used that transfer the contents of "w" to them directly. There is no way to read these registers back, so it is recommended that you keep a separate copy of these registers if you are going to use the values later.

The "w" register is an intermediate register and can be used as an accumulator. Except in the high-range devices, it is not directly addressable but can be manipulated by various instructions. Either the source register or the "w" register can be the destination of all arithmetic instructions.

The high-end PICs, while behaving in a similar manner, are a bit more complex. This complexity comes from two sources. The first is the set of primary (or "p") unbanked registers that are a subset of the complete register set. Going along with this is the location of the WREG register (which is the high-end's version of "w"), which is not separate from the other registers (and can be accessed and written to like other registers).

The high-end PIC's basic architecture looks like Fig. 2-8.

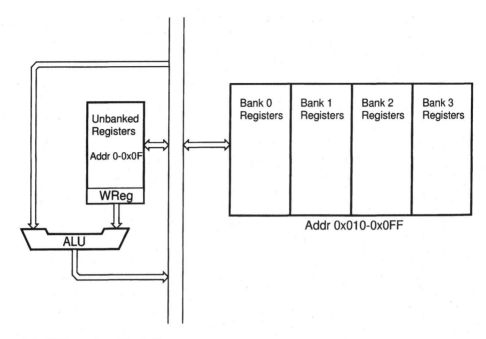

2-8 17C4x register block diagram.

Interrupts

Most new users of the PIC shy away from developing their applications with interrupts. The perception is that interrupts are difficult to master and aren't worth the effort. In this book, there are a lot of interrupt examples and explanations for a variety of situations. I find interrupts to be one of the most useful features of the PIC architecture, and they reduce the complexity of the application code.

After reading this, you might be asking yourself, "What is an interrupt?" An interrupt is a hardware event that stops execution of the mainline software to respond to the hardware event using software and then returns execution back to the mainline where the interrupt took place. See Fig. 2-9.

As you go through the book, the actual method of setting up and handling interrupts will be explained and shown.

Timers

The normal execution of a program in a processor cannot be used for timing an application. This is because different instructions take a different number of cycles to execute. The situation can be further complicated by calling subroutines or allowing interrupts to take place during the timed operation.

To eliminate this problem, timers are built into the PIC hardware.

The best way to visualize a timer is as a counter with parallel input and output that is accessible by the internal processor bus.

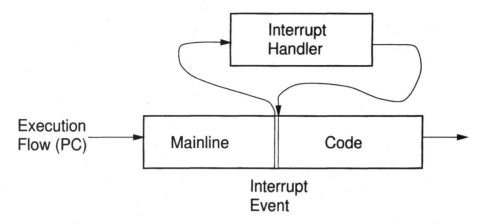

2-9 Interrupt execution flow.

Typically, a timer can use a constant clock source; however, for different applications, an asynchronous source can be used. The example of doing this that comes to my mind is the test that an engineer I worked with developed for a PC network adapter card. The output of the card was quite a complex protocol, requiring many months of effort to develop a method for monitoring it and writing a test for it. The solution used was to hook up a counter to the network output of the card and count the number of pulses that occurred in a given space of time. The test was quick to set up and surprisingly accurate in terms of being able to sort good product from bad.

In different types of PICs, there are a number of timers available. The most basic timer is known as Timer 0 or TMR0. This is an 8-bit counter that can cause an interrupt when the value inside the counter overflows (reaches 0x0100).

There are two special-use timers (TMR1 and TMR2) available in some types of PICS. These timers work in a similar manner to TMR0, but also are designed for specialized operations. These operations include doing a pulse width capture and compare and a pulse width modulated (PWM) signal. These additional timer features are located in some hardware known as the "Capture/Compare/PWM" module. In this book, I have included the descriptions of the CCP module with Timer1 and Timer2 because the CCP cannot work without them.

The last type of timer used in the PIC is the watchdog timer. It is used to reset the PIC after a specified length of time has gone by. The watchdog timer (WDT) can also be used to restart the PIC after it has been put to "sleep."

Digital I/O

Every type of PIC is able to control and read digital signals. The signals are read from and written to via the Port registers. The direction of the signals ("1" for read, "0" for write) is controlled by the TRIS registers. There is a corresponding TRIS register for each Port register. The number of bits available for digital I/O is dependent on the PIC device as well as how it's used and how many pins are required for other interfacing chores. See Fig. 2-10.

Digital input can also generate interrupts based on pin status. This feature allows the application to respond to the input conditions without having to poll the bit(s).

Parallel, digital I/O as described is the simplest way of interfacing the PIC to the outside world. The more advanced I/O functions available in hardware in some PICs can generally be simulated in software and use lower level (read "cheaper") PICs using the I/O pins only capable of digital I/O.

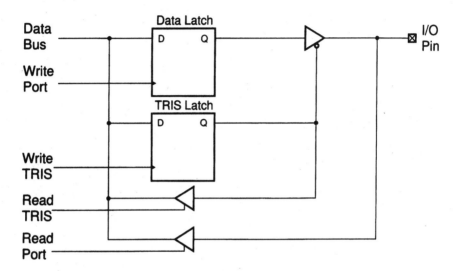

2-10 Full I/O pin block diagram.

Serial digital I/O

On some PICs, hardware has been provided to allow the PIC to communicate serially with other devices. Serial communications is a way of saying that data is shifted out one bit at a time from the PIC, rather than in parallel or all bits at the same time. Serial communications can simplify hardware connections between devices.

While the serial interfaces described here can be emulated in software on the PICs that don't have the hardware, this section will give you an idea of how the built-in serial communications hardware works.

Synchronous serial communications

The Synchronous Serial Port (SSP) module on the PIC allows the transfer of data along with a clock, used to indicate when the data is valid. The data and clock streams look like Fig. 2-11.

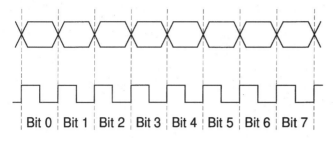

2-11
Synchronous serial data.

Bit 0 ┊ Bit 1 ┊ Bit 2 ┊ Bit 3 ┊ Bit 4 ┊ Bit 5 ┊ Bit 6 ┊ Bit 7

Typically, as is shown in Fig. 2-11, a falling edge of the clock is used to indicate when the data is valid. In the PIC, the clock is used to control the latching and shifting of each bit. Because the PIC contains an 8-bit processor, the data streams used are up to 8 bits long.

Synchronous communications can be with peripheral devices (such as serial RAMS, or analog-to-digital converters) or other intelligent controllers. In PICs that have serial capabilities, communication between devices can be carried out through communications protocols such as Philips' Inter-Integrated Circuit (I^2C, pronounced "I-Squared-C").

The I^2C protocol allows multiple devices to communicate over a single bus. This bus consists of a pull-up with open-collector drivers. Devices that can put instructions over the bus are known as *bus masters*, and devices that can only respond to instructions are known as *bus slaves*. Every device on the bus is given an unique address to facilitate communication between devices.

The advantage of synchronous serial communications over asynchronous serial communications is the speed at which the data can be transferred (essentially gated by the hardware). Care must be taken in developing the PIC synchronous serial communication application to ensure that the software can support the speed of the data coming through the line.

Asynchronous serial communications

Asynchronous serial communications consists of a transmitter sending data directly at a receiver without having to provide clocking information. While this is easier to understand and wire between devices than synchronous serial communication, it is actually quite a bit more complex to understand what is going on "under the covers."

The most common form of asynchronous serial communications is RS-232. This is a specification that standardizes electrical signal levels and data transmission rates between devices. The "data packet" being transmitted looks like Fig. 2-12.

Note that a "1" is actually a negative voltage, while a "0" is actually a positive voltage in RS-232. Because of the unusual (at least relative to TTL/CMOS) logic levels, special receiver and transmitter devices are used.

In this section, I'm going to primarily talk about the hardware used by some PICs to automatically receive asynchronous serial data. It should be pretty obvious that the algorithm used to transmit the data is identical to that of synchronous communication, except that the clock is not transmitted along with the data.

In the asynchronous data stream, there are four features that you should look for: the start bit, data bits, parity bit, and stop bit(s). Once the data is complete, the

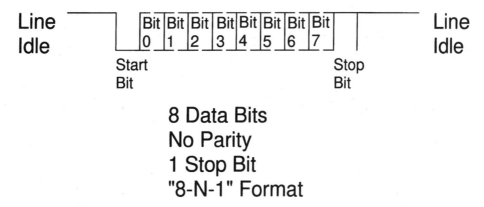

2-12 Asynchronous serial data.

transmitter goes back to sending a "1" until the next packet of data is ready to send. Each bit is active for an equal period of time (the inverse being the frequency of transmitted bits known as the *data rate*).

The start bit is used to indicate the start of the data. This bit is at a "0" logic level (also known as a *space*), and the receiver, once the level shifts from a "1" logic level (or *mark*) to a space, begins to look for data bits. Using an internal overspeed clock, the PIC hardware waits for the high to low transition and then samples it three times to ensure that is a valid start bit. Once the start bit has been authenticated (taking roughly half the time for a single bit transmission), the hardware waits a full bit transmission time (putting the clock in the middle of the first bit) and begins to sample the data coming in. See Fig. 2-13.

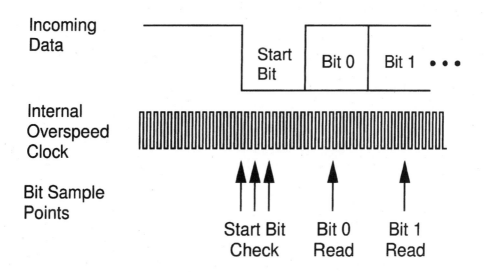

2-13 Valid asynchronous data check.

Once the device has identified the middle of the first data bit, each data bit is shifted in after waiting one period.

Once the data bits have been received (the number specified in the application), the number of 1s are totaled and then added to the parity bit. This optional bit will indicate whether or not an odd or even number of 1s should have been sent in the data bits. An *odd parity* will indicate that, when all the 1s are totalled along with the parity bit, the number will be odd (not divisible evenly by two). *Even parity* indicates that the total number of 1s transmitted will be evenly divisible by two. In the PIC, the parity bit is not generated automatically, it is left up to the software to specify the parity of the data being sent.

The last part of the information being transmitted is the stop bit(s). These bits are used by the receiver to confirm that all the transmitted data has been received and to give the receiving processor some time before the next block of information comes in. One or two stop bits can be specified.

In virtually all modern asynchronous communication application that I have seen, the data format used is "8-N-1." This means that the data to be sent is 8 bits long, no parity is used (as noted above, parity is optional), and only one stop bit is sent. This allows the most efficient transmission of the data (meaning the fewest number of overhead bits are sent with the data).

The PICs with asynchronous serial communications support also have a counter, which runs off the system clock, that is used to specify the period of each bit. While the data periods are very rarely 100% accurate, they are typically within two or three percent, ensuring the data will be received with very little chance of bit over-run or under-run (which means the wrong bit is read at the wrong time).

Analog I/O

The 16C62x and 16C7x families of PICs have analog input and output capabilities. The method used for each family of device is carried out differently. In the 16C62x family, a reference voltage is compared, while in the 16C7x family, the actual voltage is read using an "integrating ADC".

Analog input/output is a very complex topic because of the number of different ways in which it can be carried out. The period of time the analog signal is valid might also make the built-in methods inappropriate.

16C62x: *Voltage comparison*

In the 16C62x family of devices, the analog voltage comparator uses an internal reference voltage. This voltage is created from a voltage divider circuit with a multiplexer to allow selection of the actual voltage to test against it (Fig. 2-14).

For analog input, this voltage is compared against the incoming voltage; the results being binary, either VRef is greater than or less than the incoming voltage.

VRef can also be driven out of the device as an analog output.

The advantage of this way of doing analog input/output is the very fast comparisons to known values. The disadvantages include not being able to test or output values greater than Vdd.

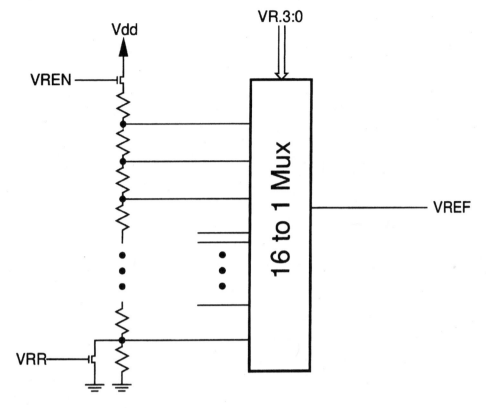

2-14 16C62x VRef circuit.

16C7x: *Analog input*

The 16C7x devices only allow reading of analog voltages. This is done by an internal analog-to-digital converter (ADC). This ADC can use either an internal voltage reference or an externally produced one to select the voltage range. The internal ADC voltage reference allows measuring voltages from 0 Volts to Vdd.

The advantages of the 16C7x analog input measurement is that the measurement is totally internal to the workings of the PIC. In the 16C62x family, if you are trying to measure a voltage, VRef has to be changed in software in a successive approximation algorithm. The 16C7x analog-to-digital converters also have the advantage that voltages greater than Vdd can be measured. The disadvantages of this method of voltage measurement include the time required to figure out what the actual voltage is.

Slaved devices

The parallel slave port turns the PIC into an intelligent 8-bit microprocessor peripheral. This clever feature allows the application designer to make the PIC look like any other peripheral attached to the data bus. See Fig. 2-15.

"Master"   "Slave"
Processor   PIC

Address ────────────────► Address
Bus   Decode ──────► _CS (RE2)

_RD ──────────────────────────────────► _RD (RE0)
_WR ──────────────────────────────────► _WR (RE1)
          Data Bus
        ◄──────────────────────────────► RD.7:0

2-15 Connection for a slave PIC.

This means that the PIC can be used instead of a custom piece of hardware. This will allow the application designer a powerful option to designing custom devices or circuits to interface with the processor's bus. It should be noted that the number of bits read/written is a maximum of eight.

Some possible uses for this feature include an intelligent keyboard controller, a custom I/O interface, or system control. The slaved device allows the application designer to turn the PIC from a small controller to a integrated part of a larger system. When using the parallel slave port, you need a method of notifying the master device when data is asynchronously available. This is usually done in the form of an unused digital I/O driving an interrupt line of the master processor.

Because the data in the slave port is not addressable (i.e., multiple addresses/ registers provided in the PIC), you will have to create a simple protocol for allowing multiple bytes to be transferred and for passing multibyte commands back and forth from the master processor to the PIC slave.

In-System Programming

All mid-range PICs can be programmed in the application circuit very simply and with minimal overhead. Programming the parts is done with a simple serial protocol. This is known as In-System Programming (ISP). As noted elsewhere, using this feature can reduce the product manufacturing costs and complexity (i.e., process steps). See Fig. 2-16.

The PIC16C5*x* and high-end PICs all use a parallel protocol that cannot be easily implemented into a larger system. Understanding which PICs use the ISP features for programming is an important consideration when picking a device to be used in an application.

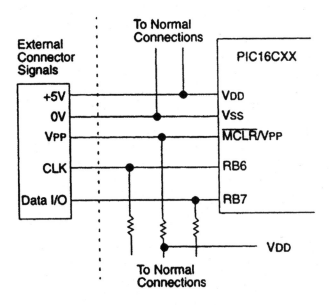

2-16 Typical in-system serial programming connection.

3
CHAPTER

The PIC processor architecture

I've always found that, when you really understand something, you can describe it in a number of different ways, in a number of different formats. At the end of this chapter, you will be able to understand the PIC architecture and how the various functional blocks interconnect.

In this chapter, as I am describing the various features of the PIC, I will also be developing a block diagram of the internal workings of the PIC.

For the 16C61, you are given a block diagram in the Microchip Datasheets that looks something like Fig. 3-1.

Looking at it for the first time can be confusing and frustrating. By going through each feature of the PIC, I hope to give you a better understanding of how each of the functional parts of the PIC interface with each other and how they work.

As I go through all the features of the PIC, I will mention hardware registers but will not give their addresses or the bits that aren't pertinent to the current discussion. You might find it useful to follow along the various sections where these registers are mentioned with a copy of the *PIC16/17 Microcontroller Data Book* from Microchip. This will give you a more complete story on how the functions are implemented, where as my description will be at a higher level.

The CPU

In the microchip documentation, you'll find that the PIC processor is described as a "RISC-like architecture . . . separate instruction and data memory (Harvard architecture)." What does this mean for those of us who don't have a PhD. in computer architectures?

The essence of the processor is the fact that the variable and I/O registers or memory are separate from the program memory. In most microprocessor architectures, variables, I/O, and programs all occupy the same execution space. This means that program data fetches have to wait for variable read/writes and I/O operations to

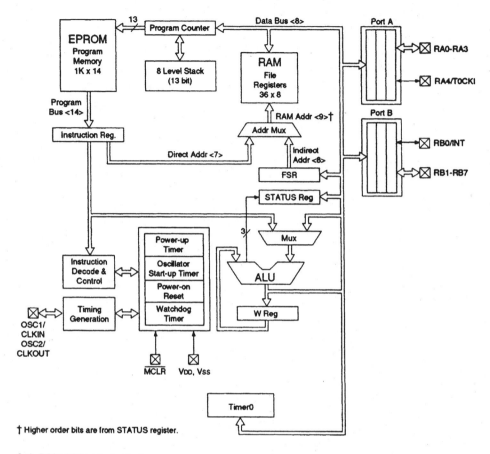

3-1 PIC16C61 block diagram.

be completed before the next instruction can be received from memory. It also means that the program memory can be overwritten by an errant program.

The PIC CPU can be thought of as an ALU feeding from and getting data to and from the various registers. There are a number of specific-use registers, which control the operation of the CPU, as well as I/O control registers and RAM registers, which can be used by the user program for variable storage. (See Fig. 3-2.)

All addresses are explicitly defined in the instructions. In the instructions, there are seven bits reserved for the addresses. This means that there are up to 128 addresses possible in the PIC.

The ALU shown in Fig. 3-2 is the "arithmetic/logic unit." This block is responsible for doing all the arithmetic and comparisons carried out in the processors instructions. Every microprocessor available today has an ALU that integrates these functions into one block of circuits.

In this discussion, there is one aspect of the PIC core that I have not discussed and that is the "w" register. The "w" register can be thought of as an accumulator or a temporary storage register. The "w" register really cannot be accessed directly, instead the contents must be moved to other registers that can be accessed directly.

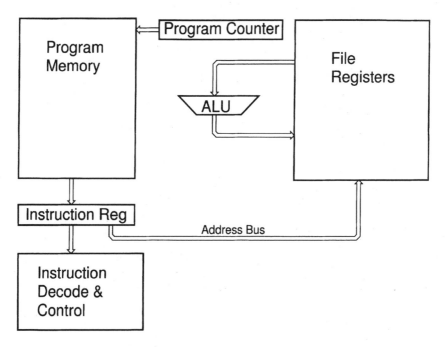

3-2 Simple architecture.

Every arithmetic operation that takes place in the PIC uses the "w" register. If you want to add the contents of two registers together, you would first move the contents of one register into "w" and then add the contents of the second to it.

The PIC architecture is very powerful from the perspective that the result of this operation can be stored either in the "w" register or in the source of the data. Storing the result back into the source effectively eliminates the need for an additional instruction for saving the result of the operation. This allows movement of results easily and efficiently.

This changes the processor diagram to the one shown in Fig. 3-3. The diagram shows the PIC at its simplest level. Well over half of the instructions can be run only using this hardware.

The last parameter of the arithmetic register operation is used to specify the destination of the result. The destination of "1" stores the result back in the source register. The convention for putting the result back into the source register is to use the character "f." Specifying a destination of "0" or "w" will put the result into the "w" register.

To specify the destination of the value, in the instruction mnemonic, the last instruction parameter is changed:

```
addwf   FSR, w    ; Add "w" to FSR and
                  ;  put result in
                  ;  "w"
iorwf   TMR0, f   ; Inclusive OR "w" with TMR0 and
                  ;  store result in TMR0
```

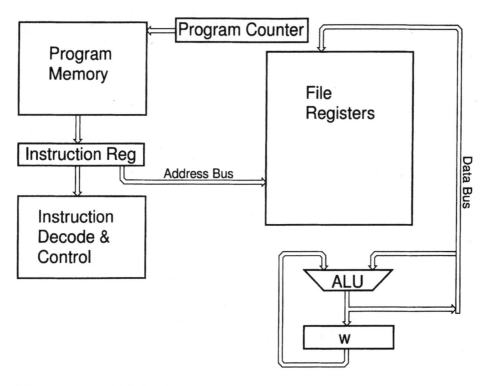

3-3 Architecture with "w" register.

Leaving the result in "w" is useful for instructions where you are comparing and don't want to change the source. It is also used in cases where the result is a intermediate value or the result is to be stored in another register.

If the destination is left off the end of the instruction:

```
addwf  Reg        ;  Add "Reg" to the contents of "w"
```

The Microchip Assembler (MPASM) and most others will interpret this to mean that the destination of the result is back into the source register (destination is "f" or "1"). Use of this format and its implications is described in more detail later in the book.

The STATUS register

The STATUS register is the primary register used for controlling the execution of the program. The actual register is broken up into three parts.

The first part is the execution status flags ("z," "dc," and "c"). These three bits are the status of the program's execution. The Zero flag ("z") is set when the result of an operation is zero (i.e., add, sub, clear, bitwise operations). The Carry flag ("c") is set when the result of an operation is greater than 255 (0x0FF) and is meant to indicate that any higher order bytes should be updated as well. The Digit Carry flag

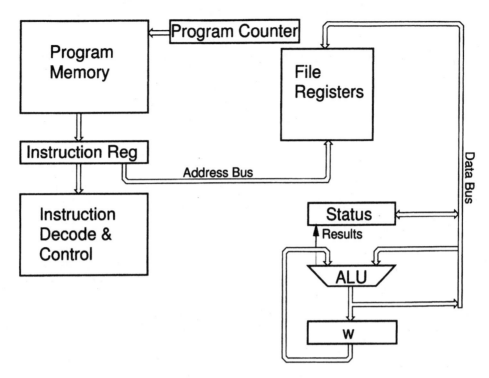

3-4 Architecture with STATUS register.

("dc") is set when the least significant nybble (four bits) of the result is greater than 15 after an arithmetic operation (add or subject). The instructions that affect these flags are listed in chapter 4 and examples of how they are changed in chapter 8

These bits can be read from or written to and are updated according to the execution of each instruction (see the end of this chapter for the explanation of each instruction and which ones are updated).

With the STATUS register storing arithmetic results, the PIC diagram now looks like Fig. 3-4.

The next two bits in the STATUS register indicate the mode in which the processor responded to Execution Start or Wake Up from sleep. Table 3-1 shows the values of the bits following different situations. The purpose of these two bits and the "PCON" register (if present) is to enable the application program to understand why the processor is at the initial program location.

Table 3-1. The _TO/_PD bits after reset

_TO	_PD	Condition
1	1	Power-on reset/MCLR reset during operation
0	1	WDT reset during normal operation
0	0	WDT wake-up from sleep
1	0	_MCLR reset during sleep or interrupt wake-up from sleep

The last set of bits are dependent on the type of processor used. For the low-end PICs (16C5x), these bits are used like the PCLATH registers in the mid-range and high-end PICs. This is described later in this chapter in the section "The Program Counter."

For the other PICs, these bits are used to indicate which bank of file/input/output/control registers can be accessed. For the PIC 16F84 and other mid-range devices, the only bit that we are concerned with is the "rp0" bit, which controls which bank of registers can be accessed directly.

The other two registers, "rp1" and "irp," are reserved for use for accessing the high two pages. While they can be read to/written from (and used as flag bits) it is *not* recommended as this might render your code unusable for future and upgraded PICs.

Figure 3-5 shows the "mid-range" register map.

In case you're wondering, the PIC16C5x series of microcontrollers don't have a second page (because these bits are used for selecting different address "pages" as opposed to file register banks). Having only one bank in the devices also means that the OPTION and TRIS registers cannot be accessed directly by the processor. To allow access to these registers, the "OPTION" and "TRIS" instructions have been included, which allow you to write the contents of the "w" register into these registers.

To new users of the PIC series processors, changing file register banks can be a confusing aspect of the PIC architecture. This is complicated by the way addresses

	Bank 0	Bank 1		
0x000	INDF	INDF	0x080	Notes:
0x001	TMR0	OPTION	0x081	• For Higher Function
0x002	PCL	PCL	0x082	PICs, the File
0x003	STATUS	STATUS	0x083	Registers may start
0x004	FSR	FSR	0x084	at 0x020
0x005	PORTA	TRISA	0x085	• The Darkly Shaded
0x006	PORTB	TRISB	0x086	Area can be used for
0x007			0x087	either additional Port
				Registers or Special
				Function Registers
0x00A	PCLATH	PCLATH	0x08A	• The Lightly Shaded
0x00B	INTCON	INTCON	0x08B	Area Denotes
0x00C			0x08C	unused Regsiters
	File Registers	File Registers		and will return 0x000
				when Read
				• The File Registers in
				Bank 1 may be
				Shadowed from
				Bank 0
0x07F			0x0FF	

3-5 Mid-range register map.

are implemented using the Microchip addressing convention. A bank 1 register has an address of 0x080 or greater (Bit 7 of the address is set, indicating page 1 is to be used) while a bank 0 register has an address of 0x07F or less. This is really just a way of writing addresses. In this format, "rp0" can be thought of as bit 7 of the address. When you see an address that has bit 7 set, before accessing the register, the "rp0" bit of the STATUS register must be set.

In chapter 8, there are examples of how to change the bank register bits as well as what can happen if it is done incorrectly.

A further complication in the page issue is that, in some of the PICs, the RAM (variable) registers are shadowed between the two pages. This means that you won't have to change "rp0" to access RAM registers in these devices if you are executing in bank 1. However, there are a number of PICs that provide separate RAM registers in both pages. This might be an issue when deciding which type of PIC and what prewritten software you are going to use for an application. To avoid this issue, I try to stay in bank 0 for variables as a default and move to bank 1 only when I have to access a hardware register or a variable array.

It should be noted that the "primary register" (STATUS, INDF, FSR, PCL, PCLATH, and INTCON) addresses are the same for each type of PIC and show up at the same address in each register page. This is designed to allow code to be used between devices without having to change register addresses when the code is used on a different device.

Register addressing

There are three different methods of getting and manipulating data in the PIC: immediate values, register addressing, and indirect addressing. Each of these modes is different and their functions overlap each other in a variety of ways, but each is designed for a specific type of operation.

Immediate addressing means that the value to be used in an operation is part of the instruction. Using the immediate operations, the "w" register can be loaded with a specific value or modified by some operation using an explicit value. (See Fig. 3-6.)

When can you use immediate addressing? Anytime you know exactly what value you want to use for an operation or load into a register.

Two examples follow. The first is setting up an I/O direction register with a specific pattern:

```
bsf     STATUS, RP0  ;  move to Bank 1
movlw   0b011011101  ;  Make bits 1 & 5 output
movwf   TRISB & 0x07F
bcf     STATUS, RP0  ;  return to Bank 0
```

Don't be confused by the ANDing TRISB explicitly with 0x07F. This is because TRISB is typically equal to 0x086, which is bigger than the page size of 0x080 (see the previous section). If TRISB was left with bit 7 of the address still set, a warning message "302" would be generated by MPASM. To avoid this, yet still have meaningful code, I clear the high bit, which specifies the page the register is used on.

The second example will do the same, but I use the 8 bit FSR register to access address 0x086.

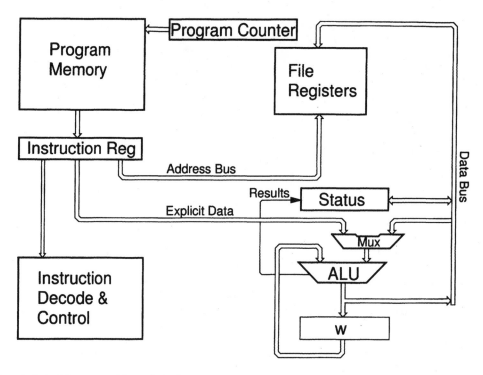

3-6 Architecture with explicit data.

```
movlw   TRISB         ; Get address of register to write to
movwf   FSR           ; Save in the FSR
movlw   0b011011101   ; Make bits 1 & 5 output
movwf   INDF          ; Store via the FSR register
```

Accessing a register means that you are processing it through the ALU and some result is placed in "w" or back in the register itself. The usual way to load the "w" register with an instruction is using the "movf" instruction. This instruction works in a very similar manner to that of the other "w" register arithmetic instructions described in chapter 4.

The following code loads the contents from a register and stores it into another one:

```
movf    Reg1, w       ; Load the contents of the first
                      ;    register
movwf   Reg2          ; Store them in the second register
```

Being able to explicitly address registers can be important, but there will be times that you want to access data indirectly and change the address arithmetically. The INDF and FSR registers are used to allow indexing or indirect addressing of data. This means that data can be kept and accessed as arrays.

The INDF Register is a "pseudo-register," which doesn't really exist. Accessing this register uses the FSR register for the address of the data you will be accessing. Our CPU diagram will now be updated to the one shown in Fig. 3-7.

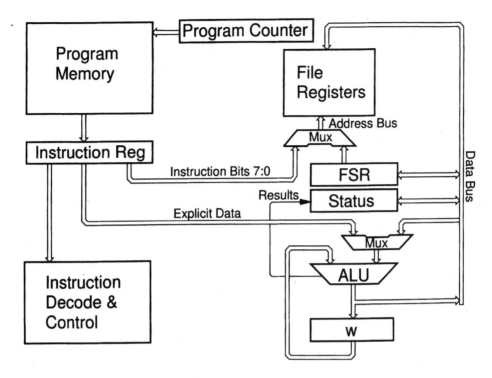

3-7 Architecture with indirect address.

For example, say you wanted to record a string of bytes coming into the PIC. This could be done explicitly with a number of variables, but this would be difficult to keep track of. Instead, a simple array of bytes could be used to allow access of the data. Individual bytes in the array could be accessed explicitly, but the whole string could be accessed arithmetically using very simple programming logic.

In C, an array is declared and accessed as:

```
char Array[ 10 ];   // Setup a 10-digit array
Array[ i ] = 'a';   // Modify an array element
j = Array[ i ];     // Read an array element
```

In PIC Assembler, this would be done as:

```
Array equ  Array_Start ; Define address for array start
                       ; Array[ i ] = 'a';
    movlw  Array       ; Modify an array element
    addwf  i, w        ; Get offset at "i" in the array
    movwf  FSR         ; Store in the FSR
    movlw  'a'         ; Get the character to put in there
    movwf  INDF        ; Store in the array
                       ; j = Array[ i ];
    movlw  Array       ; Read an array element
    addwf  i, w        ; Get offset the same way as above
    movwf  FSR
    movf   INDF, w     ; Get the character and store it
    movwf  j
```

Looking at this example, you can see that single-dimensional arrays can be implemented quite easily, as can multi-dimensional arrays. Multi-dimensional arrays are treated like single-dimensional arrays, but the index is arithmetically calculated from each parameter (i.e., the index for element [3,5] in an 8×8 array would be 3 * 8 + 5).

As noted previously, in the "baseline" PIC microcontrollers, the registers can only be accessed in one bank. This means that the "rp0" bit in the STATUS register is not used and more importantly, you cannot directly access the TRIS and OPTION registers (which are typically in bank 1 of the register space). As part of this, the most significant bit (bit 7) of the FSR is *always* set (equal to one). This means that the "incfsz" and "decfsz" instructions cannot be used on the FSR in the 16C5x PICs because the result will never be equal to zero.

There is one thing that I have to note with regards to the FSR register and indirect addressing. Even though the FSR register can access 256 different register addresses, it cannot be used to access 256 registers contiguously. The reason for this is the control registers contained at the first few addresses of each bank. If you try to wrap around a 128-byte bank address, you will screw up your microcontroller's control registers with disastrous results (an opportunity for "branching to the boonies").

Another way of looking at how the INDF and FSR registers work is to consider a "standard" processor and its Index instruction. For example, loading an accumulator from an index might use the instruction:

```
move   a, (Index)
```

The parentheses around "Index" indicate that the register contains the address of the register who's contents we want to put in the accumulator ("a").

In the PIC, rather than using INDF, we can think of the equivalent instruction being:

```
movf   (FSR), w
```

So, instead of using "(FSR)," INDF is substituted and carries out exactly the same action (the contents of FSR is used as the address of the register to use as the source).

I realize that all this is hard to understand. The first programming experiment in chapter 8 is to go through the addressing and examine the various different modes.

The Program Counter

Changing the Program Counter can be one of the most bewildering things you will have to learn with the PIC. Looking across the different family of devices, implementing "gotos" and "calls" will seem inconsistent and difficult to understand. Actually, these instructions work according to a similar philosophy, and once you understand it, they really won't seem all that scary.

In all PIC devices, instructions take one word or address. This is part of the RISC philosophy that is used for the design. This might mean that there are not sufficient bits in a "goto" or "call" instruction for the entire address of the new location of the Program Counter. Typically, all the least significant bits are put in the instruction. The most significant bits are loaded from another register, called PCLATH.

So a "goto" or "call" typically gets it's address from the instruction and the PCLATH register, as shown in Fig. 3-8.

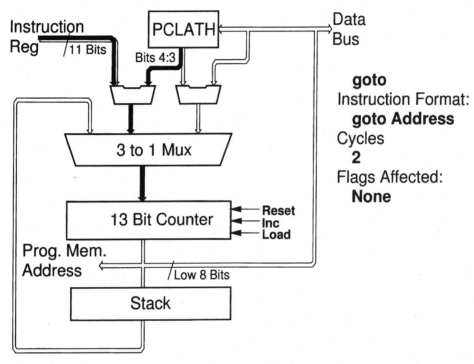

3-8 "goto" instruction.

While this form is used for all "gotos" and "calls," the method in how this works varies considerably between the three different PIC families. Later in this chapter, I go through each of the families to explain how the "goto" and "call" instructions work.

However, before I do that, I want to talk briefly about another method of changing the Program Counter. The lower 8 bits of the Program Counter are directly accessible by software and can be used to implement simple tables (described elsewhere in this book). The PCLATH register is used to provide the address of the most significant bits. This means that, before a computed "goto" can be executed, the high bits of the address (meaning PCLATH) must be valid.

Changing the Program Counter is not as simple in the PIC as it is in other processors you are probably familiar with. However, with an understanding of how the hardware works, executing "gotos" and "calls" can be done efficiently and without a lot of worrying about how they should be executed properly.

The low-end 16C5x family

The PIC 16C5x family of devices deviates the most from the model shown so far in this chapter. That is because the 16C5x does not have a PCLATH register. Instead, the high value bits are stored where the "rp" bits would be in the high-end device's STATUS register. These bits are combined with the bits in the instruction to form the Program Counter, as shown in Fig. 3-9.

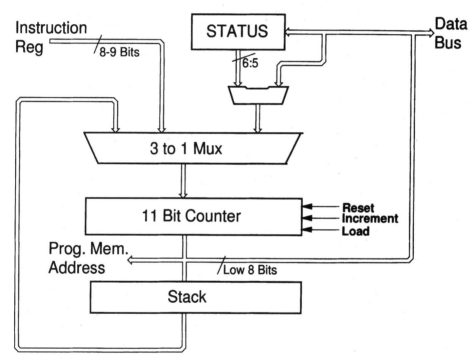

3-9 PIC16C5x Program Counter.

From the diagram, it can be seen that bits 5 and 6 of the STATUS register (the "pa0" and "pa1" bits) are used to make up the high two bits of the final address that is loaded into the Program Counter. The 16C5x "goto" and "call" instructions only provide 9 and 8 bits, respectively, of the addressing space. This means that directly, "goto" can access a 512-address "goto space" and "call" can access a 256-address "call space" (from 0 to 0x0FF). The 512-address "goto space" is usually referred to an *address page*.

To go beyond this (i.e., for the 16C56 part that has 1K addresses), the "pa0" and "pa1" bits must be set to the correct 512-address page you want the code to execute in. Once this is done, a "goto," "call," or computed "goto" is done with the PIC execution changing to the expected 512-address segment.

Computed "gotos" work exactly the same as in the other PICs with one important difference. The table must be within the first 256 addresses of a 512-address page. This means that in a 1K 16C5x, there are only two "goto" table segments that can be used, whereas in the mid-range and high-end PICs, there isn't this limitation.

Although these features might make the execution-changing code of the 16C5x parts seem a bit more complex, it really isn't that difficult to understand. In chapter 8, I show how the PCLATH register is used to control where programs are executing and by following how the code is written, you will get a good idea of how to do page jumps in the 16C5x.

The mid-range 16Cxx family

Because the mid-range PICs have a 14-bit instruction length, a much larger code space can be implemented in them. Because of this, the few simple bits in the STATUS register could not be used to follow the general rules outlined earlier quite closely. So, the PCLATH register was created to provide this function.

The Program Counter hardware of the mid-range PIC can be thought of as resembling Fig. 3-10.

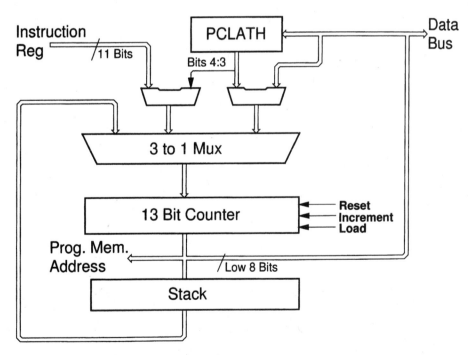

3-10 PIC16Cxx Program Counter.

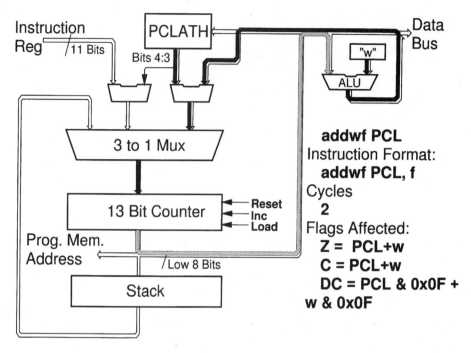

3-11 "addwf pcl,f" instruction.

With the 14-bit instruction size, up to 11 bits are specified in the "goto" and "call" instructions for the address. This means that the "page" size of the mid-range PICs is 2K.

If you look through the list of PIC devices, you will see that most have 2K or less of control store, which means that modifying PCLATH will not be required except for tables that are not in the first 256 instructions. While this is true for most cases, this is not true for computed gotos.

A "computed goto" executes the "addwf PCL[, f]" instruction directly in the PIC and can only access up to 255 addresses in a table. However, unlike the low-end parts, a table can be located at any address within the mid-range PIC. (See Fig. 3-11.)

While a table size of 255 seems to be the maximum, there are some tricks that you can do that will increase the size significantly.

The high-end 17C4x family

The high-range PICs, while similar in operation to the mid-range, differ in one important aspect, parts of PCLATH are updated during a goto/call operation.

In chapter 6, I discuss the idea of setting up functional blocks of code, where all the values use the same PCLATH. This is obviously going to be a problem for the high-end devices when tables are used within the functional block. The easiest way to make sure this isn't a problem is to make sure that PCLATH is reloaded before each table read.

Peripherals

In the previous figures, I haven't shown how the peripheral registers are set up in the PIC. In terms of blocks, the whole register set can be thought of as being architected exactly the same as a regular computer.

This could be thought of as being very analogous to how peripherals are wired to your PC's "ISA" bus.

The address bus is the bus shown as the short line in the architecture diagrams. Each peripheral is actually a block that is hung on the address decode logic, data bus, and controlled by the instruction decode hardware. So, to create a new design with a new feature, the chip designer drops a new device onto the data bus and runs lines from the address decode and instruction decode logic. I'm simplifying this a lot, but this is essentially what is done.

Interrupts

Interrupts are, sadly, one of the most unused features of the PIC. I guess this stems from the notion that interrupts are hard to understand and implement. Nothing could be further from the truth. While being quite easy to implement, interrupts give you a whole new world in terms of PIC application flexibility. With a properly written interrupt handler, you'll find that your application mainline will actually become easier to write and debug.

An interrupt occurring during code execution can be thought of graphically as resembling Fig. 3-12.

In this section, I'm not going to focus in on any specific PIC device or the instructions required to set up and process the interrupt. This will be found for the PIC16C84 in chapters 8 and 9. Setting up interrupts and handling them is unique for each type of interrupt. This is not to say that handling interrupts is difficult—it's actually quite easy. What I'm trying to say is that each type of interrupt is handled dif-

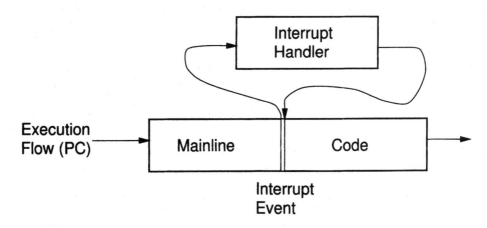

3-12 Interrupt execution flow.

ferently, although the process for understanding how to handle them is the same regardless of the type of interrupt.

In all PICs, the INTCON register (located at address 0x0B in the PIC memory map) is the main source of interrupt control (See Fig. 3-13). The most significant bit ("Global Interrupt Enable" or "GIE") is used to allow interrupts to take place. If this bit is reset, then interrupts cannot take place. Upon interrupt entry, GIE is reset by the hardware before the interrupt handler begins execution. When the interrupt handler has finished, the GIE bit is set by the return from interrupt instruction. It is not recommended that you change GIE inside the interrupt handler, because this might lead to unexpected execution within the PIC by way of nested interrupts.

The next most significant bits are the interrupt-enable flags. These bits are set to allow interrupts when their respective interrupt functions occur. As a pointer, in the PIC documentation, it is very easy to find the Interrupt Enable bits; the labels that are given to them all end in "IE." Along with the Interrupt Enable flags in "INTCON" are the Interrupt Active flags, which are set when the expected interrupt conditions are met. The interrupt active flags (which end in "If") can be set even if the respective Interrupt Enable flag or GIE is reset. It is also very important to note that these bits must be reset before a return from interrupt ("retfie") instruction is exe-

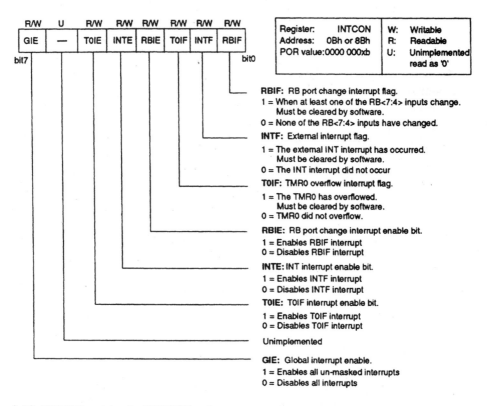

3-13 INTCON register for PIC16C61 only.

cuted. If the bits are not reset before the interrupt handler ends, then the interrupt handler will be called immediately following the "retfie."

Additional Interrupt Enable bits and Interrupt Active flags might be located in other registers throughout the PIC. Actually, this is true for pretty well all the PICs. This means that before setting GIE, you have to make sure that all the Interrupt Enable flags are in the correct state. This is made considerably easier by noting that all Interrupt Enable bits are reset upon power-up, and to enable an interrupt, the IE bit must be set.

To explain the sequence of events for programming and handling interrupts, I've put in the following explanation:

```
1. Initialize the interrupt source hardware.
2. Enable interrupts—Set the GIE and IE of the appropriate interrupt in
   INTCON.
3. Wait for the Interrupt to happen—This could be waiting for an
   interrupt in a loop or executing normally and processing the data
   from the interrupt when required.
4. Interrupt—Wait for the current instruction to complete.
5. Handle the interrupt request—If the IE flag is not set for actual
   interrupt, go to step 8.
6. Execute the interrupt handler—Start execution at address 4.
7. Reset the interrupt controller and return from the interrupt.
   a. Reset the Interrupt Active flag
   b. Set the expected IEs for the next interrupt.
   c. Execute the "retfie" instruction.
8. Resume executing at the instruction following "interrupted" one. Go
   to 3.
```

It is important to note that more than one source can have an interrupt pending at any time. Different interrupts can be handled by checking the appropriate Interrupt Active flag (IF).

The interrupt latency period is the time it takes for the interrupt handler to start executing. The interrupt processing time is how long the program takes to handle the interrupt, reset the hardware, and return to the mainline. This value is found by using the simulator and understanding how long the program takes (basically finding it empirically). You could determine this by looking at your interrupt handler and counting the number of cycles you expect it to take, but determining the value by the simulator is the easiest and most accurate method I've found.

Note that handling the interrupts involves saving and restoring the state of the mainline program. Care must be taken to ensure that the registers containing the execution status are restored before returning from the interrupt. Also, just as importantly, the state of the interrupt handler must be set up correctly (i.e., correct page) before it can start executing. Chapter 5 describes how this is done.

Timeout errors can be handled very efficiently using TMR0 and interrupt. This is demonstrated in the infra-red receiver used in "I/R tank" in chapter 9.

Parallel input/output

The input/output pins used in the PIC are actually quite simple, consisting of a read/write port and a data output (or tristate driver) control (TRIS) register. The hardware is shown in Fig. 3-14.

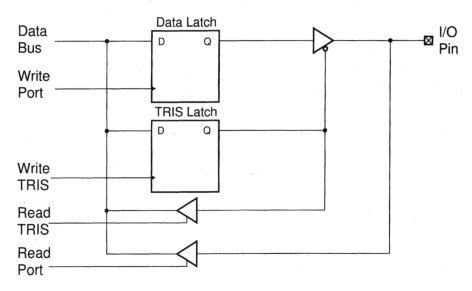

3-14 Full I/O pin block diagram.

Both the "Data" port and TRIS registers appear as standard registers within the register address space in the mid-range and high-end PICs. In the low-end PICs, the TRIS register can only be accessed via the "TRIS" instruction.

Data can be written to the output port at any time, but no bits will be output unless the corresponding TRIS register bit is reset (equal to zero). The TRIS state can be remembered as "1"nput, "0"utput.

Data is always read from the pin (not from the Output register). This can lead to some potential problems with instructions that read and then write back data to a port (i.e., "bcf" and "bsf" instructions). If a bit, who's current state is important to the execution of the program later, is actually pulled to a different level by external hardware and then a "bcf," "bsf," or "movf PORT*n*" instruction is executed (where the register is read and then the contents are put back), then the value stored in the output latch of the register will be changed to the new value. This problem is explained and investigated in great detail later in the book.

Along with the "Standard" port shown earlier, there are a number of features that are added to the parallel I/O port to enhance its functionality.

One of the most common enhancements is the use of an "open collector" (actually "open drain") pin instead of a standard CMOS output (as is shown previously). (See Fig. 3-15.)

This allows the pin to be bussed with other open collector drivers to form a "Dotted AND" bus (when one line becomes active low, the bus becomes active low).

Actually, there is a trick of simulating an open-collector driver using standard PIC drivers. The port bit is programmed with a "0" and then left in "input" mode (high impedance). This simulates a "1" in an open collector. When a "0" is to be asserted, the bit is put into output mode and the low bit is driven on the bus, pulling it down.

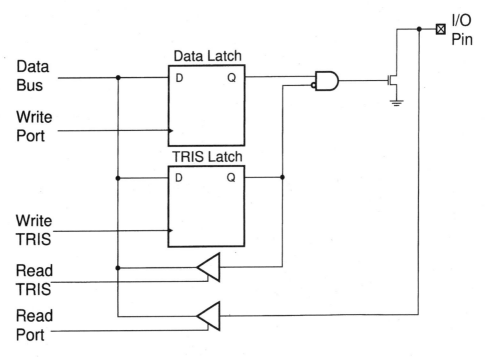

3-15 Common collector pin diagram.

Other enhancements to the parallel I/O pins include having a programmable weak pull up (similar to a 50K pull-up resistor) on the pins of PORTB. This allows the use of switches pulling down to ground without any additional parts.

There are also two interrupt request hardware types that can be added to the pins in the form of a separate Interrupt Enable pin or an Interrupt on Change in the external status of a number of input pins. Using these options will be explained later in the book.

Timers

One of the features that you will use a lot of in the PIC processors is the timer functions. Every PIC has two types of timers that are available for use by the application developer. The first, known as TMR0, is an 8-bit counter with two sources and the ability to cause an interrupt in the mid-range and up PICs. The second is the watchdog timer, which is a hardware timer that is primarily used to reset the PIC to detect software problems. Different versions of the PIC might have additional timers that are used for different functions.

TMR0 is a highly programmable 8-bit counter. The source of the counter can either be the cycle count or an external digital signal. Using the cycle count mode along with a crystal will give you a very precise and predictable clocking signal, which can be used for accurately determining event timings. The external source allows counting of external events as well as timing off of a clock. Selection of which clock source to be used is made by the RTS bit in the OPTION register.

The prescaler (which is specified in the OPTION register) is selectable between TMR0 and the watchdog timer. The prescaler is an interesting beast in that it only passes signals on to the counter after some fixed number of clock cycles. The number of cycles is determined by the value put in the least significant three bits of the OPTION register. Figure 3-16 shows what delays are possible for different control bits in the prescaler.

If the prescaler is not used, then the clock will be incremented after every two cycles of the input source. The timer circuit can be modelled as shown in Fig. 3-17.

Using the timer might not be as straightforward as you might expect. As noted earlier, the timer is incremented on a certain number of clock cycles. When the clock reaches 0x0100, the "T0IF" interrupt request flag is set (and the interrupt request handled), and the timer resets itself and continues counting.

To specify the interval between T0IF "ticks," I use the following formula:

$$Delay = \frac{(256 - InitTMR0) * Prescaler}{\left(\dfrac{Frequency}{4}\right)}$$

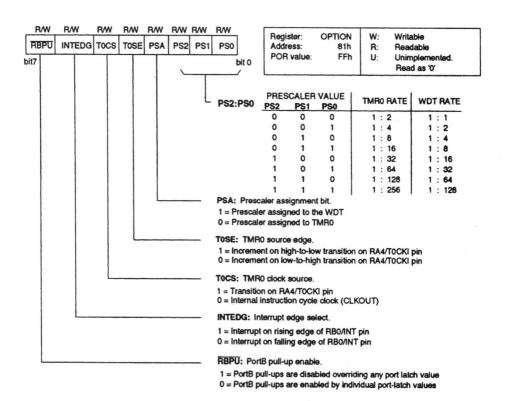

3-16 OPTION register.

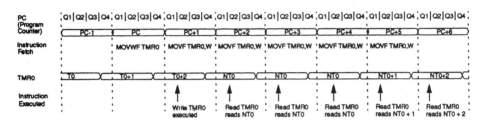

Note 1: Bits, T0SE, T0CS, PS2, PS1, PS0 and PSA are located in the OPTION register.
Note 2: The prescaler is shared with Watchdog Timer (refer to Figure 7-6 for detailed diagram)

3-17 Timer0 (TMR0) block diagram.

where:
- *Delay* is the delay in seconds.
- *InitTMR0* is the initial value put in the TMR0 register.
- *Prescaler* is the prescaler value (if no prescaler is selected, use 2).
- *Frequency* is the PIC's clock frequency in Hz.

This Formula can be manipulated to:

$$InitTMR0 = 256 - \left(\frac{Delay * Frequency}{(4 * Prescaler)} \right)$$

to provide the initial timer value for a given delay.

One point to mention about the clock is that changing the TMR0 value resets the prescaler and the 2x cycle counter before the clock. This is an important thing to remember in critical, timed programs; the act of resetting the clock might screw up your timing. If you must have a constant clock interval, you must use a clock and a prescaler value that will give you the interval (you can ignore the interrupt latency in the previous formula) you can work without having to write to the TMR0 register.

The most obvious use for the timer is as an elapsed time counter that interrupts your program to allow something that is supposed to happen. A good example of this is using the timer to wait for an event to occur and then processing. "Frosty the Snowman" in chapter 9 is an example of just this. The timer can also be used as a stopwatch, allowing you to time events and react according to what value is in the timer (see "I/R tank" in chapter 9). The timer can also be used as an error time-out value (again, see "I/R tank" in chapter 9).

The watchdog timer (also known as "WDT") is clocked by an oscillator built directly into the PIC chip. The nominal timeout period is 18 ms, which can be extended by the use of the prescaler. In the explanation of the OPTION register earlier in this chapter, the PS0 to PS2 bits showing the TMR0 prescaler delays also include the watchdog timer delays. Using the full value of the prescaler, the watchdog timer interval can be extended to 2.3 seconds.

The watchdog timer is enabled by setting a bit in the configurations fuses and cannot be disabled by software. The watchdog timer can only be reset by executing a "clrwdt" instruction. This resets the WDT and prescaler (if selected for the WDT) and restarts the count from zero.

This complicates previous diagram a bit, and the timing circuit can be represented by Fig. 3-18.

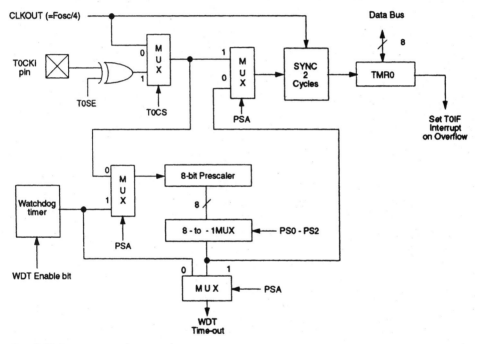

Note: T0SE, T0CS, PSA, PS0-PS2 are bits in the OPTION register.

3-18 Block diagram of the TMR0/WDT prescaler.

A WDT timeout will cause the PIC and the _TO bit of the STATUS register to be reset. If you plan on using the watchdog timer to "rein in" programs that run amok, then you should make sure that the beginning of your mainline checks the _TO bit upon Reset Entry. The watchdog timer can be used to wake up the PIC during "sleep." In this mode, an application can be put down, waking up only to check the environmental conditions, execute according to them, and go back to sleep until it's time to check the outside conditions again.

TMR1, TMR2, and the Capture/Compare/PWM (CCP) modules

Timer1, Timer2, and the CCP modules are only available in the PIC16C6x and PIC16C7x series of PIC devices. This hardware is designed to provide advanced timing features to your applications as well as the ability to interface with different types of external devices without requiring additional hardware or substantial software. While I'll discuss these devices, I do not provide any software or hardware examples for them in this book.

Timer1 executes and uses hardware that is very similar to the base timer (TMR0) except that it is 16 bits wide (TMR0 is only 8) and it has circuitry similar to that of the CPU clock, which allows you to put on an oscillator that is separate from the CPU clock. This other oscillator can be used for real-time clock crystals or can be driven from external sources. The external timer can be synchronized with the processor clock by selecting the synchronized counter mode (part of the CCP).

TMR1's operation can by controlled by the CCP module for used in measuring the length of a pulse. (See Fig. 3-19.)

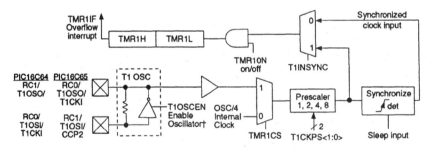

† When the T1OSCEN bit is cleared, the inverter and feedback resistor are turned off. This eliminates power drain.

3-19 TMR1 block diagram.

TMR1, when driven by an external clock source, will continue to run during sleep, if it is not being used with the CCP module. The ability to run during sleep gives you another avenue to wake up from sleep after a specified period of time.

TMR2 operates differently from TMR0 and TMR1 in that its output can be used to generate specific, repeating pulses. This *Pulse Width Modulated* waveform can be used for controlling dc motors and servos. (See Fig. 3-20.)

The value in TMR2 is continuously compared against the value in register PR2. When TMR2 is greater than PR2, a signal is asserted (the signal is de-asserted when TMR2 times out, or equals zero). This signal can be routed to a postscaler so that interrupts can occur after a specific number of cycles. (See Fig. 3-21.)

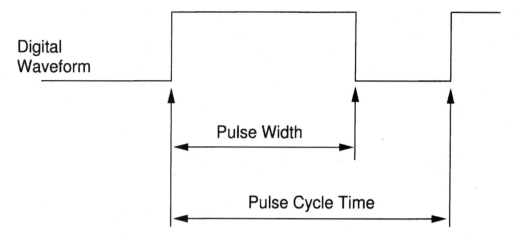

3-20 PWM pulse.

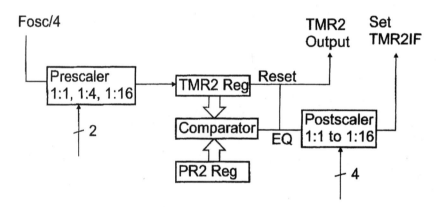

3-21 Timer2 block diagram.

The CCP module is used to provide an interface for TMR1 and TMR2 to allow them to:

- Indicate that a pulse of a specific length has been encountered ("Capture" mode with Timer1).
- Compare incoming pulses to a specific length of time.
- Output Timer2 as a Pulse Width Modulated signal for controlling dc motors and servos.

PIC input/output control registers

Parallel Slave Port

The Parallel Slave Port (PSP), as implemented in the 33 I/O pin devices, uses the Port "D" and "E" hardware. This feature turns the PIC into an intelligent device controller accessible through a standard processor data and control bus. (See Fig. 3-22.)

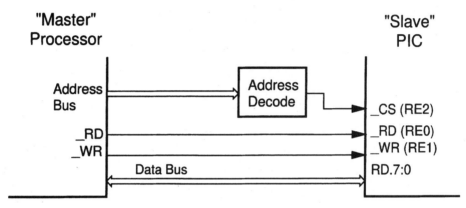

3-22 Connection for a slaved PC.

To turn on the Parallel Slave Port, the PSPMODE bit must be set (equal to 1) in TRISE along with the three Port "E" direction control (TRIS) bits. This sets up the Port "E" bits as the _Read, _Write, and _Control bits for the PSP interface to the master processor. The contents of the TRISD register becomes irrelevant when PSP is enabled.

The hardware inside the PIC looks like Fig. 3-23. As you can see in this diagram, interrupts can be generated when the host microprocessor accesses the PSP Port (either read or write). This interrupt is enabled by the PSPIE in the PIE1 register, and the active bit is in the P1R1 register. Along with interrupts, the status of the data transfer between the PIC and the master processor can be monitored by polling the _RD, _WR, _CS, or interrupt status bits.

By using the Parallel Slave Port feature of the PIC, you can transform a standard PIC into a virtual, intelligent microprocessor peripheral.

Serial I/O

Serial communications hardware is built into a number of the PIC devices. The serial communications hardware devices are known as the Synchronous Serial Port (SSP) module and the Serial Communications Interface (SCI) module. Both modules behave differently but provide substantially the same features.

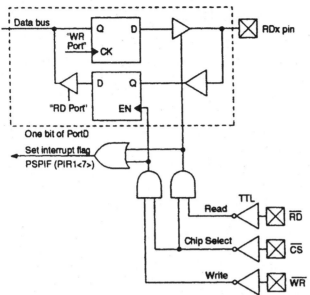

3-23
Summary of parallel slave
port registers.

Note: I/O pins have protection diodes to VDD and VSS

Synchronous serial communication

Synchronous serial data transmission can be loosely defined as transmitting data when the receiver expects it. To ensure that the receiver is synchronized with the transmitter a common clock (controlled by the transmitter) is used to latch in the data. Synchronous Serial Communications can be carried out in both the SSP and SCI modules. A synchronous data stream looks like Fig. 3-24.

In the SSP module, transmitting and receiving data is quite simple by setting up the appropriate bits in the SSPCON and SSPSTAT registers. Data is received and transmitted through the SSPBUF register. This is known as the Serial Peripheral Interface (SPI) mode and is best suited for communicating with 8-bit serial devices. (See Fig. 3-25.)

With synchronous serial devices, multiple receivers can be put on the same data and clock lines. This complicates transmission considerably. The PIC SSP module resolves these issues in two different ways.

The simplest method is by use of the serial slave bit. When this bit is set low, the SSP is enabled to receive data at all times. The software controlling the PIC then decides whether or not to transmit to the next device in the chain. This means the software developer must create an arbitration scheme between devices so that the appropriate devices transmit and reply.

The second method of allowing multiple receivers on the same bus is to address each of the receivers and develop a protocol for this communication. There is built-in hardware in the SCI Module that supports the Inter Integrated Circuit ("I²C" also known as "I-Squared-C") communication.

This protocol is designed to allow multiple devices to communicate over the same bus. The I²C hardware in the PIC is primarily responsible for notifying the pro-

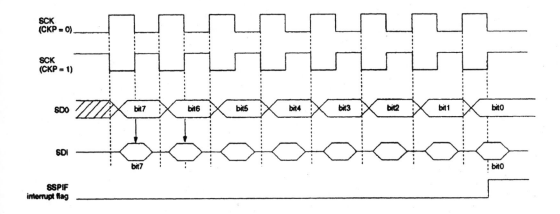

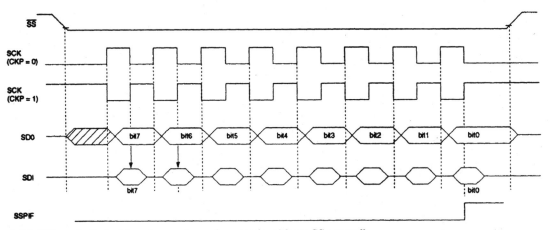

3-24 SPI mode timing (master mode or slave mode without SS control).

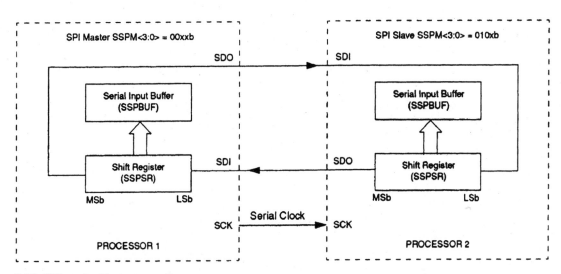

3-25 SPI master/slave connection.

gram that its address has been received on the bus. It is also responsible for putting acknowledgements on the bus for messages sent to the PIC. (See Fig. 3-26.)

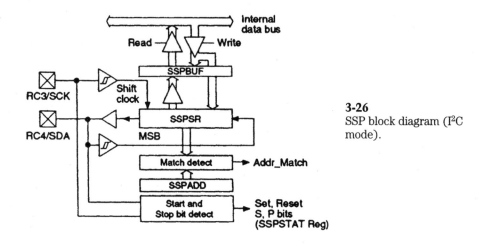

3-26
SSP block diagram (I²C mode).

This means an unique address must be programmed into the PIC before any I²C communication can take place. The address can be 7- or 10-bits long and is obviously specific to the specific network the PIC is in. The network address is often programmed in manually by the use of a DIP switch or hardcoded into the software.

Unlike many of the other hardware interfaces described here, the I²C interface requires quite a bit of software to support it. This also means that there are a number of different I²C protocols in use. For this reason, it is important that you understand the protocol used in the target application before programming the PIC.

There is one very important aspect of I²C that should be noted. I²C transmitters (drivers) are open collector drivers. This allows multiple drivers on the bus.

The PIC provides the necessary open-collector drivers for I²C along with the SSP module. If you are going to implement an I²C protocol using a PIC without an SSP Module, it is important for you to use the output pins that are open-collector output.

Asynchronous serial communication

Asynchronous communications can be loosely defined as the ability to transmit data without a synchronizing clock. Serial data looks like Fig. 3-27.

This definition is actually quite inaccurate because the receiver actually *does* synch up to the transmitter. This is done in the SCI module by waiting for the start bit to go low, then finding the middle of the start bit and then polling the data bits afterward. (See Fig. 3-28.)

As shown in Fig. 3-28, there is an overspeed clock that is used to find the middle of each bit. Along with this, there are three samples taken for each bit, with a majority detector (which indicates the level of two or more samples). This helps to ensure that the asynchronous receive is accurate.

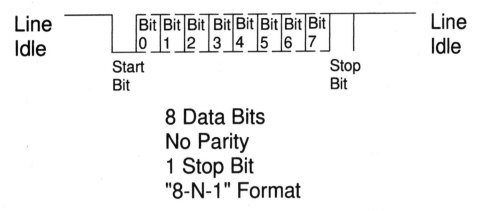

3-27 Asynchronous serial data.

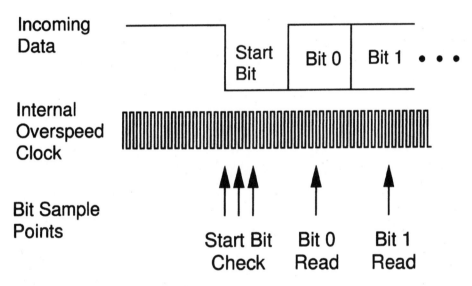

3-28 Valid asynchronous data check.

Transmission of asynchronous data is very similar to that of a synchronous data transmit. The major difference is with regards to the start, parity, and stop bits. (See Fig. 3-29.)

There are three important side issues in regard to asynchronous communication. They are the data rate, parity, and handshaking. These side issues complicate asynchronous communications, but not so much that they can't be overcome.

The data rate (colloquially known as "baud rate") is the frequency at which data bits are transmitted. Some people get extremely obnoxious and will give you lectures on the "baud rate" versus "data rate" versus "bit rate." When I describe the data rate, I am describing the number of bits (including the overhead of start, parity, and stop) that can be transmitted in a given period of time.

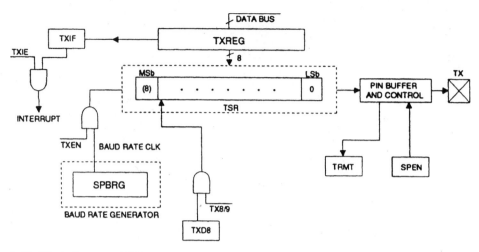

3-29 Block diagram of SCI transmit.

There are a number data rates that are most common. Most equipment that you will connect to will use these values (in bits per second, or bps):

- 300
- 1200
- 2400
- 4800
- 9600
- 14,400
- 19,200
- 28,800
- 38,400
- 57,600
- 115,200

If these seem familiar to you, it's because most modems (which are asynchronous serial devices) use these data rates.

Generating these data rates requires some care with the hardware Baud Rate Generator in the PIC. You might want to tailor the PIC's clock speed to the final data rate to minimize the transmitting/receiving data rate error to a minimum. The SCI module's Baud Rate Generator can work in two modes: high and low. The mode selection is simply made by a bit (called BRGH) that changes a clock divisor between 16 and 64.

The formula for determining the data rate is:

$$Data\ rate = \frac{PIC\ clock\ frequency}{((16 + (48 * !BRGH)) * (x + 1))}$$

where x is the value programmed into the Baud Rate Generator.

To determine what value for x to use to get the desired data rate, the formula is rearranged into:

$$x = \frac{PIC\ clock\ frequency}{(Data\ rate * (16 + (48 * !BRGH))) - 1}$$

This means that, when BRGH is equal to zero, the Baud Rate Generator is set to the low range; otherwise, the Baud Rate Generator is in the high range.

When I said that the PIC clock frequency could be dependent on the speed you want to transmit serial data at, I was speaking about the *gross* level. The clock frequency selection mostly affects the range of possible data rates.

Asynchronous communication is quite tolerant of speed differences between the transmitter and the receiver. When determining the baud rate, you can easily calculate the difference between the actual frequency and the desired frequency. From this, you can determine whether or not the data transmission will be reliable (i.e., bits moving relative to the RX clock to the point where there is an opportunity for a bit to be read in the wrong position). This value is typically greater than 7% before bits begin to be read at the wrong location. This means you can get reliable asynchronous communications with nonprecision devices such as ceramic resonators (with typically 1% to 5% tolerance). RC clocking schemes do not have the appropriate level of accuracy for asynchronous communication.

Parity and data length are two issues to be considered in the PIC. As I've pointed out elsewhere, most modern serial asynchronous communication is in an 8-N-1 format, which means that 8 data bits are transmitted with a start, a single stop, and no parity information. You might have to communicate with some devices that have different data lengths or parity schemes.

Different data lengths are possible in the PIC with some software overhead. To change the data length from 8 bits to few bits, simply putting in 1s for the high bits (which aren't used) will simulate a shorter data word.

Parity is not actually calculated in the PIC hardware. This optional bit is a simple form of error detection. This is something that you will have to do and put in the high bit of the transmitted word. The PIC has the ability to transmit and receive up to 9 bits. This gives you the capability to transmit up to 8 bits of data with a parity bit. As noted previously, fewer bits can be transmitted by placing 1s after the parity (which is the highest value) bit.

RS-232 (a standard that uses the asynchronous protocol) specifies the use of a number of "handshaking" lines for controlling the flow of information (i.e., transmitting when the receiver is ready). These lines can be implemented simply in the PIC using digital input/output lines.

Providing the handshake lines can burn up a lot of effort and control store space. Whenever possible, I try to avoid using these lines and implement what is known as "three-wire RS-232." The three wires are the Transmit, Receive, and Ground. With the high speeds and superior signal integrity of modern computers, the handshaking lines are not required for reliable data transmission and reception.

If there are times you don't want to have data transmitted to the PIC, you can implement a "XON-XOFF" protocol. The "XON-XOFF" protocol simply transmits control characters from one device to another, to tell the second device whether or not it can transmit data to the first device. The "XON-XOFF" protocol is very simple to implement.

Analog input/output

There are two different methods used by PICs for internal analog voltage measurement. These methods have advantages over each other and best suited for different applications.

The PIC16C62*x* family compares an input voltage to an internal reference voltage. This comparison can take place very quickly. The internal reference voltage can also be output for use as an analog control.

The PIC16C7*x* devices use a capacitive charging method to measure the input voltage. This method actually returns a value for the voltage (the 16C62*x* just returns a binary value indicating whether or not the input voltage was higher or lower than the internal reference). There are two advantages to this method of analog measurement. The first is that the PIC hardware will figure out the value of the voltage without any software overhead (unlike the 16C62*x*, which would require you to develop an algorithm to figure out what the voltage actually is). Secondly, it can measure voltages greater than Vcc.

The methods used by the two devices can be simulated very easily with external hardware.

The PIC 16C62x *Analog Input/Output module*

The 16C62*x* simply provides two analog voltage comparators wired to various input pins of the PIC. It can also provide an analog voltage output. The analog comparators and their output look like Fig. 3-30.

The result of the comparison is a binary value (1 if Vin is greater than VRef). When this value changes, an interrupt can be generated within the hardware.

The control registers for the output of the comparators is stored in a the CMCOM register. Comparator configurations are set up by using bits in this register.

The comparator multiplexors are controlled by the "CIS" bit of CMCOM. This selects (where appropriate) the input to the comparator.

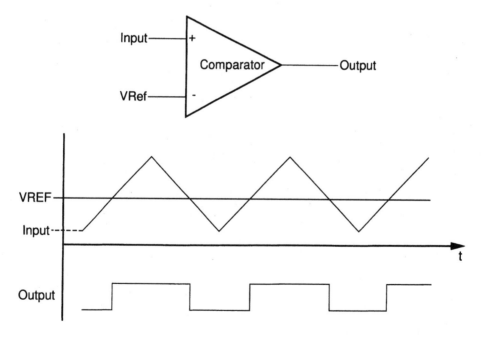

3-30 16C62*x* comparator operation.

An interrupt can be generated when either one of the comparator outputs changes state. The interrupt request is initiated by setting standard interrupt control bits (located in INTCON, PIE1, and PIE2). The comparators will continue to run if the PIC is put to "sleep" and the interrupts are enabled.

The VRef circuit (which provides the analog voltage for the comparators) is also very simple. Vref is produced by a 16-stage resistor ladder. (See Fig. 3-31.)

Note that VRef can be produced in two different voltage ranges as defined by the following two formulas. If the VRR bit equals 1:

$$\text{VRef} = \left(\frac{\text{VR.3:0}}{24} \right) * \text{Vdd}$$

otherwise:

$$\text{VRef} = \left(\frac{\text{Vdd}}{4} \right) + \left(\frac{\text{VR.3:0}}{32} \right) * \text{Vdd}$$

This means that VRef falls in the following two ranges. If VRR equals 1:

$$0 >= \text{VRef} >= \text{Vdd} * \frac{2}{3}$$

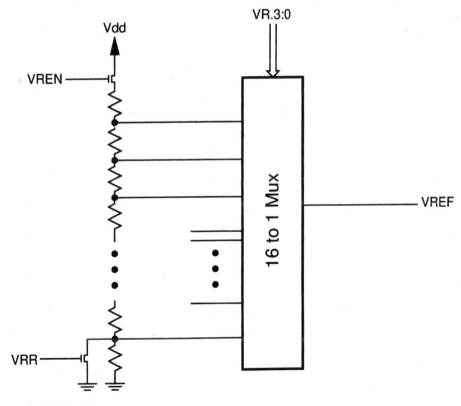

3-31 16C62*x* VRef circuit.

otherwise:

$$\frac{\text{Vdd}}{4} >= \text{VRef} >= \text{Vdd} * \frac{3}{4}$$

VRef can be output to the RA2 pin for analog voltage output. If this pin is used for output, then it is recommended that a unity gain buffer be used to supply drive current to external circuitry.

PIC 16C7x *Analog Input module*

The 16C7x devices actually have a built-in analog-to-digital converter (ADC). This "module" consists of a capacitor that is allowed to charge and then compared using a binary search algorithm that begins when the "Go/_Done" bit is set:

```
ADCChk = 0x080              - Start with the high value
ADRES = 0                   - A/D value

while ADDCChk != 0
  ADRES = ADRES + ADCChk    - Set voltage to current point
  if ADRES > VInput         - If the check value is greater than input
    ADRES = ADRES - ADCChk  - then change the value
  ADCCheck = ADCCheck > 1   - Shift down the check value
endwhile

Go/_Done = 0                - Reset bit to indicate that ADC is done
```

This algorithm could be reproduced in software to provide an ADC function for the 16C6x.

The result is found to 8-bit precision. The 16C7x ADC can run during sleep and can be used to wake the device up out of sleep as well as interrupt the currently executing code. This check can be interrupted or shortened. However, for the analog-to-digital converter to be as accurate as possible, it must be allowed to run as long as required to ensure an accurate measurement.

The equivalent analog input circuit is shown in Fig. 3-32.

Because the length of time required for the A/D operation is variable, the software should poll of the "Go/_Done" bit or wait for a measurement complete interrupt to be requested.

The input pin used is software selectable through an analog multiplexor.

CPU Program Counter stack

The CPU Program Counter stack is a simple way of storing where the program was executing when a subroutine is called or an interrupt is serviced. The classic analogy used for an execution stack is a stack of trays in a cafeteria. When you're about to do (take) something, you take a tray off the stack, when you are finished, you return the tray to the stack.

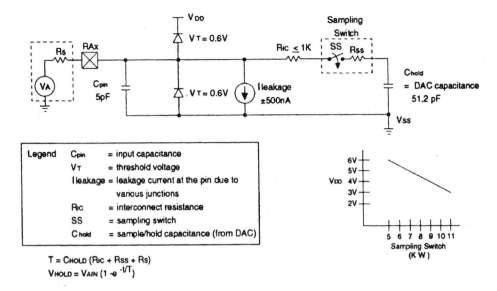

3-32 Analog input model.

The PIC's execution stack works in exactly the same manner. When a program is going off somewhere for a while (a subroutine or interrupt), a new Program Counter is set up, and the old one is saved on the stack (or, following the previous analogy, on the tray that was picked up). When execution is returned, the Program Counter is restored to its previous value. (The Program Counter is taken off the tray before the tray is returned to the stack.)

The PIC Stack does work differently from typical microprocessor and microcontroller architectures. Normally, registers and miscellaneous data can be stored on the stack as well as the Program Counter.

This is not possible with the PIC microcontrollers. Only the Program Counter can be put on and off the stack. The stack operations are implicit and the stack cannot be manipulated directly. This means that care must be taken in ensuring that the stack is accessed in the correct manner.

The stack is also only 8 locations deep in the mid-range PICs and above and only 2 locations deep in the low-range PICs (the $16C5x$ series of devices). This means that care must be taken to ensure that no more than 8 calls can be done with the PIC. Actually, this is not an unreasonable limit if your software is *not* recursive.

If you've had any experience with any other microprocessors or microcontrollers, you probably feel that the PIC cannot compete with them in terms of data passing on the stack during subroutines. Actually, that isn't true. The PIC can use the FSR and INDF registers as a stack very efficiently. For example, say you wanted to write a routine that caused a specified delay. In stack operations, this might look like:

```
push    dlay
call    dlay_routine
pop     nothing
```

In the PIC, this could be simulated as:

```
movf   dlay, w
movwf  INDF
incf   4,1
call   dlay_routine
decf   4,1
```

While an instruction or two extra is executed to implement a stack over other processor architectures, the FSR can be used as a very efficient stack pointer. A programming example of how to implement this is demonstrated in chapter 8.

Sleep

When the PIC isn't going to be used for a period of time, you might want to put it into a low-power mode called "sleep." This is done when the PIC isn't required to execute for a while and when it will be required to execute again is known well beforehand.

"Sleep" is at it's lowest possible power consumption when the I/O pins are set to nonchanging values (to prevent unwanted current flow, which results in increased power requirements).

During "sleep," if the PIC is driving the oscillator, it is turned off. The current state of all the registers is saved. What doesn't happen is that the timers (except for TMR1, if it has an external clock source) continue to run. The watchdog timer does continue to run, however. The name "sleep" for the state really isn't a misnomer; the device is really not working until it is woken up in some way.

Waking up from sleep can be done a number of different ways, including toggling the reset pin (_MCLR), external interrupts, and the watchdog timer. Timer1, if available, can be used to wake the PIC up from sleep, but the other internal timers cannot (simply because they are not running). Depending on how the PIC is woken up, the program will start at a different location.

One important thing to remember about sleep is how long it will take the PIC to get out of it. During sleep, all the major functions of the PIC are shut down (except for the watchdog timer), and this includes the oscillator. To ensure that the oscillator is stable after wake-up, 1024 oscillations will be counted before the PIC begins to execute again. This delay must be taken into account in your application. The PIC sleep/wake cycle looks like Fig. 3-33.

Toggling _MCLR or waiting for a watchdog timer time-out will result in the PIC being reset and starting to execute at address 0. Since address 0 is also the address the PIC starts up at on initial power-up, the "_TO" and "_PD" bits, along with the "PCON" register bits (if available), must be checked to understand why the program is at the reset address. Using registers for this function is not recommended because, at initial power up, they are at indeterminate values, which might match the test values.

Table 3-1 gives the _TO/_PD bits after reset for different conditions.

External interrupts (pins or states) can be used to take the PIC out of sleep as well. There is one interesting aspect of this that should be noted. If the GIE bit is set,

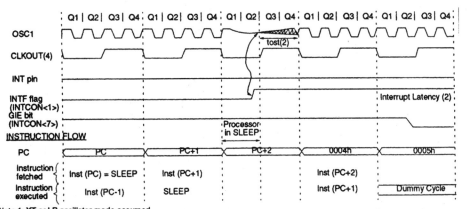

3-33 Wake-up from sleep through interrupt.

then the PIC will reset back to the interrupt handler address (4). If it's not, the PIC will continue executing the program where it stopped to go to sleep.

Actually, I lied. After an external interrupt, the PIC will execute the instruction *after* the "sleep" before doing anything else (i.e., jumping to the interrupt handler at address 4). Wake up from an interrupt is similar to Fig. 3-33; however, after the first instruction after sleep is executed, the interrupt handler is executed.

Because the instruction after sleep is always executed, it is recommended to always put in a "nop" instruction following the sleep instruction. By doing this, you won't have to think about how the PIC will be operating after wake up from sleep.

Configuration fuses

The configuration fuses are used to define power-up conditions of the PIC. These conditions must be programmed into the PIC to ensure that it will power up in a state required by the application. The configuration fuses cannot be read by the PIC itself and reside at program address 0x02007.

Configuration fuses control such things as the oscillator type to be used (resistor/capacitor network, crystal, and the frequency range, or an external oscillator), turning on the watchdog timer, using the self power-on reset (PWRT), and the code protect (which prevents the code from being read from the PIC). In the code supplied in this book, you will see a variety of different combinations used and how they affect the PIC. I should point out that the values for the configuration fuses do *not* have to be set in the program; they can be supplied to the programmer before the device is programmed. However, I do not recommend this method because it involves having to specify the fuse values each time a device is programmed. As I first started using the PIC, this caused some problems because I didn't use the same values between program iterations (making it much harder to debug the application) and causing a bit of head scratching when the PIC refuses to run).

There is one word of caution when setting the configuration fuse values. Do *not* use the code-protect bit unless you are absolutely sure. In new PICs, setting this bit locks out any future access of the control store (writing as well as reading). The bit is protected from U-V erasure in the EPROM devices. This means that the device can only be used for the given program and nothing else. If you have to change the code after the code-protect bit is set, you will have to take the PIC out of the application and put in a new one. While there are cases where this is desirable, caution should be taken before setting the bit.

The "ID bits" can also be considered to be part of the configuration fuse discussion. These bits, residing at instruction addresses 0x02000 to 0x02003, can be used to indicate the code version level, device serial number, or whatever is required by the application.

While all the bits can be read/written by the programmer, typically only the least significant 4 bits are used. This means you have a total of 16 bits available for this application.

Like the configuration fuses, these bits cannot be accessed by the processor. Generally, I let the programmer software determine a CRC value to be put in the ID bit location. This allows me to check version levels easily.

The configuration fuses and ID bits are accessed in MPASM using the __CONFIG and __IDLOCS pseudo-instructions, respectively. For the configuration fuses, the possible values are defined in the PIC ".inc" files included with MPASM. They are referenced by the actual PIC name (i.e., the ".inc" file for the 16C84 is p16c84.inc).

PIC 17C4x *architecture*

The high-end PICs differ from the low-end and mid-range devices in two major respects. The first is the device's ability to interface via a microprocessor-like external bus. The second (and probably more significant) difference is the inclusion of the "w" register in the register bank. Both of these features make the 17C4x devices different from the other PICs in terms of software development and capabilities.

External addressing

The high-end PICs can run in three different memory modes. (See Fig. 3-34.)

The memory mode that the PIC runs is selected by using the configuration fuses (specified by the "__CONFIG" assembler instruction). These different modes give the 17C4x quite a bit of flexibility in regard to how the PIC is used. Regardless of the memory mode used, the PIC register locations do not change.

The microcontroller mode means that the PIC runs exactly the same way as the other PICs, the control store is located inside the PIC, and no pins are used for interfacing. There is only one device-protect mode and that works with the microcontroller mode (the memory modes using external memory obviously can't take advantage of this mode with external memory devices).

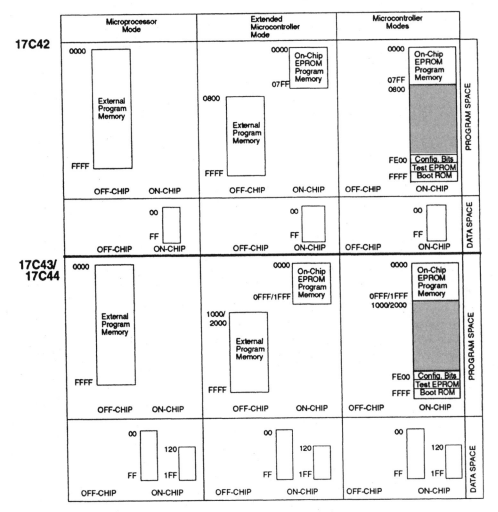

3-34 Memory map in different modes.

The extended microcontroller and microprocessor memory modes use I/O bits in the C, D, and E ports. These bits form a multiplexed address/data bus along with control signals. The typical method of wiring external devices is by using an address latch as shown in Fig. 3-35.

There are a few important things to note about this mode. The first is that the address space is either less than 64K words (in extended microcontroller mode) or 64K (in microprocessor mode). This means the PIC can actually access 128K bytes! Each byte can be read from or written to by using the "table" commands. Because a write capability is included in the table commands, RAM can be used to provide extra storage space to the PIC.

Along with memory devices, it can be seen that other devices can be attached to the external memory bus. This means that typical microprocessor interface devices

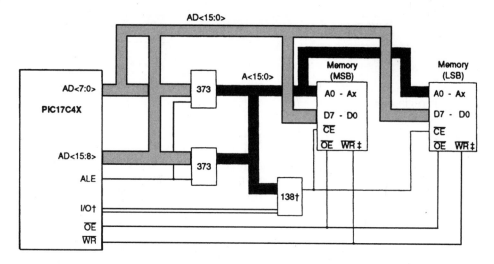

† Use of I/O pins is only required for paged memory.
‡ This signal is unused for ROM and EPROM devices.

3-35 Typical external program memory connection diagram.

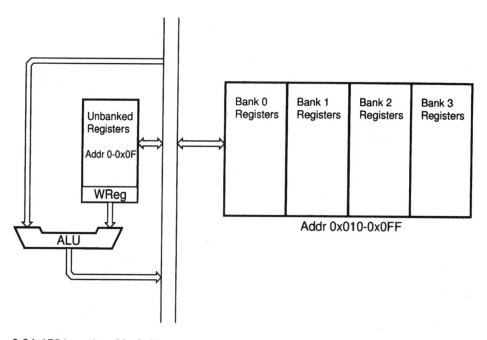

3-36 17C4x register block diagram.

can be used with the PIC as "memory-mapped I/O." They are accessed using the table read/write instructions.

Registers

The high-end devices use a different register model. (See Fig. 3-36.)

This register model means that the PIC can transfer data to the execution registers much faster and more efficiently than in the mid- and low-range devices. If you were to take a program that appears in this book and assembled it for a 17C4x device, you'd find many errors, mostly centered around the use of the "movf" instruction (which doesn't exist for the 17C4x family).

This instruction doesn't exist in the high-end devices because of the ability to transfer data back and forth between the "p" (unbanked) and "f" (banked) register set.

While this seems like a real blessing, there is one problem that you must be aware of. The high-end instruction set doesn't support the use of the "movf Reg, f" (which checks the value of "Reg" and sets the zero flag accordingly). This cannot be done in the high-end devices unless the register to be checked is in the primary set. This means that it might not be possible to save a value in "w" and check another register in the high-end processors. This is probably not a big problem for most applications, but it is something the programmer has to be aware of.

The unused (by the current code) registers in the primary set can be used for temporary storage of variables. This feature can be used for passing data between banks, or saving data in available registers irregardless of which bank is active.

4
CHAPTER

The PIC instruction set

To characterize a microprocessor's instruction set, I find that it is best to break the instructions into functional groups. The PIC instruction set can be broken up into four such groups.

The first group contains the arithmetic instructions, which are used to read or write the contents of file registers. This includes adding and subtracting from registers, along with incrementing, decrementing, and doing bitwise operations. The arithmetic instruction group can be broken up into two subgroups: the register arithmetic (where only the contents of registers are used) and the immediate arithmetic (where an explicitly stated constant value is used for the operation).

Execution-control instructions make up the next functional group. These are the "gotos," "calls," and "returns" as well as conditional instruction skips. The PIC instruction set differs from other traditional instructions in that a "jump on condition" requires two instructions instead of a single explicit one.

To carry out conditional jumps or other conditional operations, a skip next instruction is executed before the actual operation. Conditional execution is shown in this chapter as well as throughout chapters 8 and 9.

The next group is the microprocessor control group and consists of processor-specific instructions. These instructions primarily affect the operation of the processor and the hardware associated with it. In this group, the "sleep" command (which puts the PIC into a state of suspended animation) and the watchdog timer clear (which resets the watchdog timer) are the main instructions in this group. This group also includes the "option" and "tris" instructions, which are used to write directly to hardware control registers in bank 1.

The last group is the register bit-manipulation commands (set bit and reset bit). I've put these instructions into their own group because they are unique to the PIC architecture. These bits allow direct manipulation of individual bits in registers. The most obvious use for these commands are for direct control of microcontroller hardware, but they do have some advantages that I will explain later.

All these instructions are common to the various PIC families, but there are some unique instructions for the 17C4*x* (high-end) architecture. These instructions are explained at the end of the chapter.

Register arithmetic instructions

Register arithmetic operations (also refer to as "byte-orientated file registers operations" by Microchip) are the instructions used in the PIC to transfer data between registers as well as carry out mathematical operations on the data within the registers. This is a $10 way of saying that these instructions move data around in the PIC as well as combine it arithmetically with other data.

If you first look through the PIC instruction set, you will probably be very surprised to see how limited it is (30 plus instructions are listed for the low-end and mid-range devices). However, once you have experience with it, I think you'll be impressed with how powerful it is, and you will see that you are able to carry out a wide range of operations. Part of this flexibility is the ability to specify operation destinations. This means that the PIC can often carry out operations in single instructions that would require two or more in traditional processors.

Part of the flexibility of the instruction set is the different ways in which registers can be accessed. The instructions described in this section can read and write all the registers in the PIC. Register addresses are specified in the instruction itself as a 7-bit number. This means that the data can only be accessed within the current bank ("rp0" is used to specify the current bank). Data can also be accessed using the FSR (index) register by accessing INDF pseudo-register (at address 0). The PIC is unusual in that indexed addressing uses exactly the same format as direct addressing: the register address specified activates the indirection hardware.

"movf" instruction

The "movf" instruction is used to set the Zero flag according to the contents of the register and can be used to load the "w" register with the contents of the specified register. (See Fig. 4-1.)

This might seem like a weird way of describing the instruction, but it is actually quite accurate in what the instruction does. Like the other register arithmetic operations, the result of "movf" can be stored in the "w" register.

However, if the "movf" instruction is specified with putting the source back into its register (i.e., a "movf reg, f" instruction), what is it actually doing?

It's loading up the value from the register, checking it in the ALU to see if it is equal to zero, and then putting the value back. So, in this way, I think of the primary function of "movf" as being used to set the Zero flag. The secondary function of the instruction (because this is optional) is to load "w" with the contents of the source register.

Register clear instructions

The PIC processors have instructions that can be used to explicitly clear various registers. The "clrf reg" instruction places zero in the specified register. "clrw" sets the Zero flag of the STATUS register.

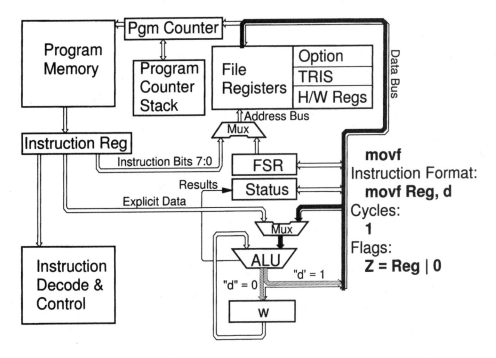

4-1 "movf" instruction.

One thing to remember about the clearing and "movf" instructions is that they also set the Zero flag. This might cause problems for you, if you are trying to maintain a status for later program execution. For this reason, you should use only the carry flag as a passed status between routines, because it is affected by fewer instructions than the Zero flag. (See Fig. 4-2.)

"movwf" instruction

The "movwf" instruction is used to store the contents of the "w" register into the specified file register. If the register is INDF, then the destination will be the address pointed to by the FSR register. No status bits are affected during the execution of the instruction. (See Fig. 4-3.)

"addwf" instruction

The arithmetic operation that probably first comes to mind is addition. In the PIC, addition is carried out in a very straightforward manner, with the result produced as expected. (See Fig. 4-4.)

It should be noted that all the operation status bits are affected by the operation. The Zero flag is set if the result ANDed with 0x0FF is equal to zero. The carry flag is set if the result is greater than 0x0FF (255).

The Digit Carry flag is set when the sum of the least significant four bits (also called a "nybble") is greater than 0x0F (15).

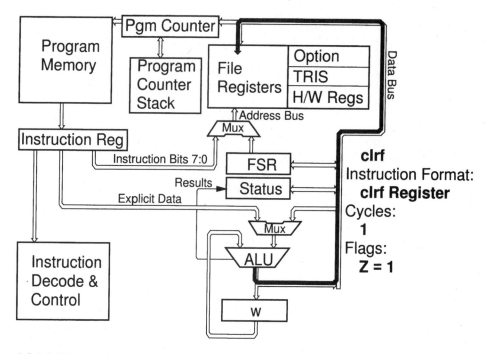

4-2 "clrf" instruction.

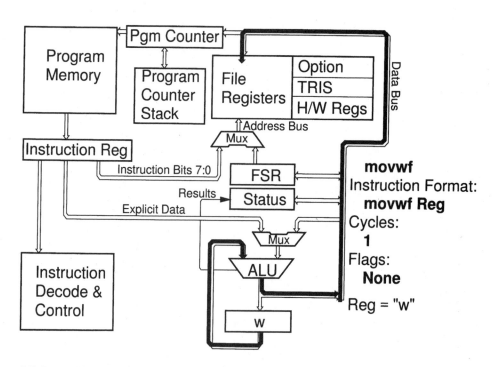

4-3 "movwf" instruction.

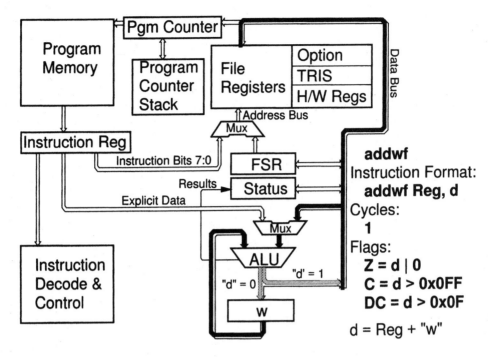

4-4 "addwf" instruction.

For example, if you had the code:

```
movlw  10          ; Add 0x00A to 0x00A
movwf  Reg
addwf  Reg, w      ; Put the result in "w"
```

the "w" register would contain 20, "Reg" would still have "w," the zero and carry flags would be reset (equal to zero), and the Digit Carry flag would be set.

"subwf" instruction

Subtraction in the PIC is something that you should look over and understand thoroughly before you use it. (See Fig. 4-5.)

Looking at the information in Fig. 4-5, you'll probably be completely confused. The best way to explain subtraction in the PIC is to note that it is not subtraction at all. Instead, it is adding a negative value to the source.

So, instead of "subwf" being:

```
Destination = Source - "w"
```

it is actually:

```
Destination = Source + (-"w")
```

The negative "w" term of the previous equation is done by substituting the two's complement negation formula:

```
Negative = (Positive ^ 0x0FF) + 1
```

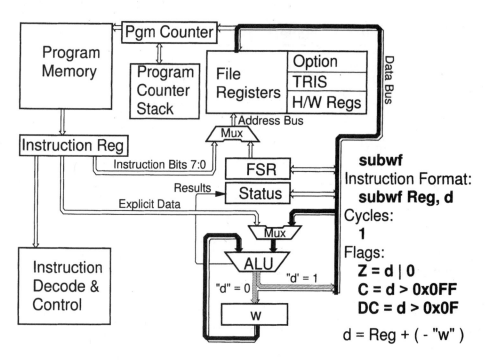

4-5 "subwf" instruction.

This means that the subtraction formula now becomes:

```
Destination = Source + ("w" ^ 0x0FF) + 1
```

I find that, when I am using the instruction, it helps to remember this formula, because I can easily understand what "subwf" is doing and predict how it will behave.

Remembering this formula also helps me to understand how the carry flags work. Looking at the instruction, the Carry and Digit Carry flags probably run counter-intuitively to what you expect (and might have experienced with other processors).

For example, what happens when you subtract 2 from 1 in the PIC:

```
Source = 1
"w" = 2
Instruction = subwf Source, w
```

To try to understand it, we use the previous formula and plug in the values for "Source" and "w," getting:

```
w = 1 + (2 ^ 0x0FF) + 1
```

Following it through, we get:

```
w = 1 + ( 0x0FD ) + 1
w = 1 + 0x0FE
w = 0x0FF
```

which is what we expect. Note, however, that the Carry flag would not be set, which is *not* what we expect in a typical processor. If a subtract instruction was actually executed, we would expect that the Carry flag would be set in this example (typically "carry" becomes "borrow" in other microprocessors).

Now, for the same example, where:

```
Source = 2
w = 1
```

we can follow the formula:

```
w = 2 + (1 ^ 0x0FF) + 1
w = 2 + (0x0FE) + 1
w = 2 + 0x0FF
w = 0x0101
```

0x001 (0x0101 & 0x0FF) is actually stored in "w." However, note that, in this case of subtracting a lower value from a higher value, the Carry flag (and, possibly, the Digit Carry flag) is actually set!

For this reason, after a "subwf" or "sublw" or adding a negative number, I like to refer to the Carry flag as the "Positive" flag.

If you look back at the explanation of the instruction, it should now make a lot more sense.

Bitwise logical operations: "andwf," "iorwf," and "xorwf"

The bitwise operations—"andwf," "iorwf," and "xorwf"—allow you to do the basic logical operations on the bits of a register. Probably the "and" and "xor" operations will be familiar to you, but Microchip has elected to call the "or" as *inclusive or* (i.e., "ior"). This works the same way as the traditional "or" you are familiar with. (See Fig. 4-6.)

The contents of the register specified are operated on with the contents of the "w" register. The Zero flag in the status register is set or reset depending on the condition of data following the operation.

If you wanted to check the contents of a register, you could load "w" with zero and then XOR it with the contents of the register to check.

For example, jumping to a specific location if PORTB is equal to 0x0A5 would be accomplished by:

```
movlw   0x0A5      ; Get the Check Value
xorwf   PORTB, w   ; XOR it with the Expected Value
btfsc   STATUS, Z  ; Do we Have a Match?
goto    PORTB_A5   ; Yes, Execute the Specific Code
```

"comf" instruction

The "comf" instruction (Fig. 4-7) is used for inverting all the bits in the source register. Following a bit in the footsteps of the "subwf" instruction, I want to note that this instruction does *not* negate a number (in two's complement format).

A negative can be produced using the "comf" instruction knowing that:

```
Negative = (Positive ^ 0x0FF) + 1
```

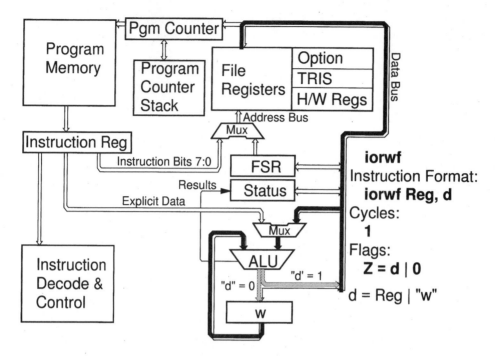

iorwf
Instruction Format:
iorwf Reg, d
Cycles:
1
Flags:
Z = d | 0

d = Reg | "w"

4-6 "iorwf" instruction.

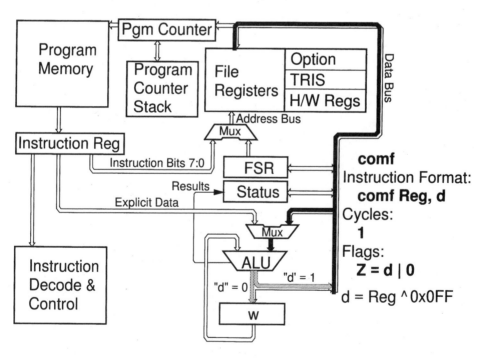

comf
Instruction Format:
comf Reg, d
Cycles:
1
Flags:
Z = d | 0

d = Reg ^ 0x0FF

4-7 "comf" instruction.

The "comf" instruction is equivalent to XORing a value with 0x0FF. So to negative a file register:

```
comf    Reg
incf    Reg
```

If the result should be in "w" and the source should not touched, the code would be:

```
comf    Reg, w
addlw   1
```

Note that this sequence will only work with PICs that are able to execute the "addlw" instruction (namely the mid-range and high-end PICs).

"swapf" instruction

The "swapf" instruction exchanges the nybbles in a file register. As with the other instructions in this section, the destination of this exchange can either be the "w" register or the source register itself. What truly makes this instruction special, however, is the fact that none of the status flags (Carry, Digit Carry, or Zero) are affected. This last feature makes "swapf" useful in some circumstances. (See Fig. 4-8.)

From strictly a data movement point of view, the "swapf" instruction can be used for two purposes. The first is to allow the application program to store two digits in a single file register, swapping between the digits depending on which one you want

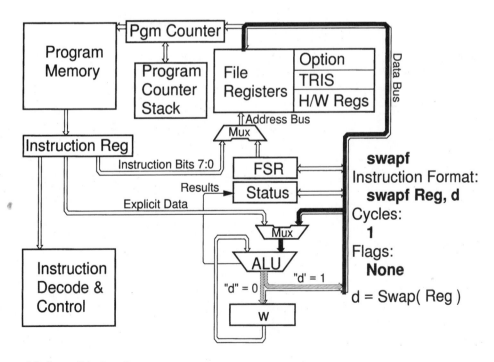

4-8 "swapf" instruction.

to access. The second is to do a fast four-position shift (either left or right depending on which nybble you AND with 0x0F).

I often use "swapf" to separate a byte into two nybbles (4 bits each) for displaying. For example, to output a byte in hexadecimal format, the following code would be used:

```
swapf   Byte, w      ; Get the High Nybble
andlw   0x00F
call    PrintHex     ; Print It out
movf    Byte, w
andlw   0x00F
call    PrintHex     ; Print the Low Nybble
```

One of the most useful things the "swapf" instruction can do for you is make sure that status registers are not changed when loading "w." This takes advantage of the fact that "swapf" does not change the current status register contents after execution. (This is typically used in the restoring of the context registers before returning from an interrupt.)

Rotate instructions: "rlf" and "rrf"

The rotate instructions are useful for a number of reasons. The basic function of rotate is to move a register one bit upward or downward, with the least significant value being loaded from the Carry flag, and the most significant value put into the Carry flag. (See Fig. 4-9.)

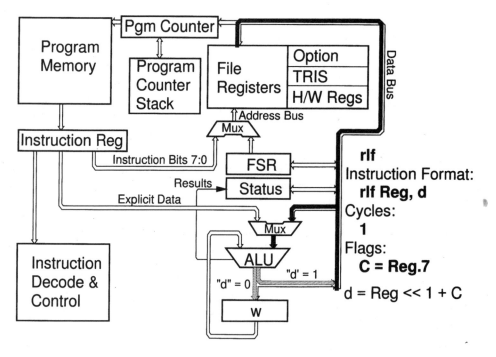

4-9 "rlf" instruction.

The rotate instructions can be used for doing multiplication and division on a value with powers of two. This can also be done on 16-bit values. The following example shows how to multiply a 16-bit number by 4:

```
bcf   STATUS, C    ; Clear the Carry flag before rotating
rlf   Reg, f       ; Shift the value over (multiply by 2)
rlf   Reg + 1, f
bcf   STATUS, C    ; Now, repeat to multiply by 2 again
rlf   Reg, f
rlf   Reg, f
```

Another use is to shift data in/out of a register serially (this will be seen later when data is shifted in/out of various registers from external devices).

Increment/decrement instructions: "incf" and "decf"

Increment ("incf") and decrement ("decf") are used to change the value in a register by 1. (See Fig. 4-10.)

Following the completion of an increment/decrement, only the Zero flag is changed. You might have expected the Carry flag to change state (if the value goes over 0x0FF for "incf" or below 0 for "decf"), but that isn't the case.

This means that incrementing or decrementing a 16-bit number isn't quite as straightforward as you might expect. Actually, I should change that: The 16-bit decrement isn't as straightforward as you would expect.

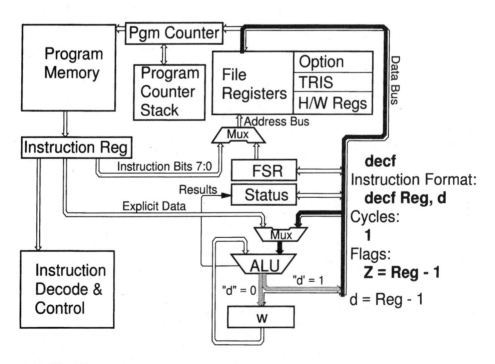

4-10 "decf" instruction.

The 16-bit increment can be done by:

```
incf   Reg, f       ; Increment the low byte
btfsc  STATUS, Z    ; Are we at 0? (Low = 256)
 incf   Reg + 1, f  ; Yes, increment the high
```

Because the zero bit is set when the low byte is equal to 0 (or 256, or any multiple of 256), we know when to increment the high byte of the 16-bit number.

The decrement isn't quite so simple. If a value reaches zero during a decrement, then all you've got is zero, and zero does *not* indicate that the high byte should be decremented. Therefore, you must actually use an instruction that changes the Carry flag after the low-byte decrement to ensure that you know when to decrement the high byte. This is done by subtracting one from the low byte, rather than decrementing it, like so:

```
movlw  1            ; Load "w" register with sub value
subwf  Reg, f       ; Subtract it from the low byte
btfss  STATUS, C    ; Is the carry bit set?
 decf   Reg + 1, f  ; No-we rolled over, decrement high byte
```

Add/subtract one and skip if result equals zero: "incfsz" and "decfsz"

Along with the bit test instructions (described later in this chapter), there are two other instructions that skip on a given instruction. They are the increment/decrement skip if the result is equal to zero. (See Fig. 4-11.)

These two instructions work exactly the same as the "incf" and "decf" instructions in terms of data processing. One is added or subtracted from the source register. This value is then stored either in "w" or back in the source register. The important difference in these instructions is, if the result is equal to zero following the increment/decrement, the next instruction is skipped.

This means that "decfsz" and "incfsz" can be used for loop-control operations. Actually, I should say that "decfsz" is normally used for loop control. The following code example shows how a loop can be repeated 37 times with very little software overhead:

```
movlw  37            ; Load the Count register
movwf  LoopCounter
Loop                 ; Repeat for each iteration of the loop
  .
  :
decfsz LoopCounter, f ; Decrement the Count register
 goto  Loop          ; If not == zero, loop again
                     ; Continue on with the program
```

This code can be used anywhere a loop is required and, as you can see, the overhead is only four instructions.

While "incfsz" is not as readily usable as "decfsz," it can be used to create a very "tight" timing loop to get a 16-bit delay value:

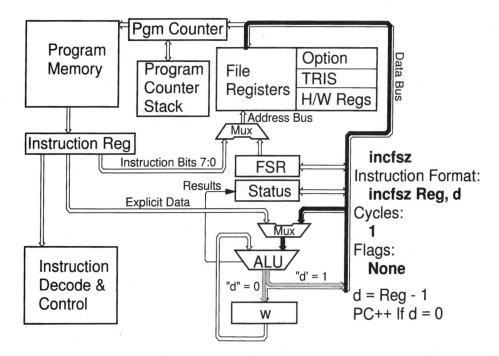

4-11 "incfsz" instruction.

```
Loop                    ; Timing loop return
  incfsz Count          ; Increment the least significant byte
    incf  Counthi       ; Inc most sig byte if least
                          != 0
  btfsc  PORTn, Bitn ; Jump out of loop if conditions are met
   goto  Loop
   movf  Counthi, w   ; Now, get correct value for most sig
  subwf  Count, w     ; byte of the count
  movwf  Counthi
```

In "Loop," the least significant byte of the count is incremented each time through. The most significant byte is incremented if the result of incrementing the least significant byte does not equal zero.

This means that the most significant byte is equal to the count of the least significant byte minus the value of the most significant byte (which can be thought of as having the value of the least significant byte minus the most significant byte).

This little snippet of code only uses five instruction cycles (or 20 clock cycles) and is the minimum 16-bit timing code that you can produce for timing an event not using a timer. One (potentially serious) drawback of using this block of code is that there is no way to escape if there is an overrun (i.e., the event doesn't end in $64K \times 5$ instruction cycles). Despite this, this loop is an extremely efficient method of providing 16-bit event timing with only five instruction cycle granularity (or possible error).

A couple of notes on these two instructions. If you are using them on processor registers, care should be taken to ensure that the hardware registers are capable of reaching 0. In the PIC16C5x (low-end) series of PICs, some registers contain bits that can never be 0 (are always equal to 1). These registers will never reach 0, which means that they would never skip the next instruction.

As well, these instructions do not affect any status flags (Zero would probably be expected). This means that you might want to put a "bsf STATUS, Z" after the instruction following the incfsz/decfsz instruction.

For example, in a loop:

```
decfsz Count       ; Decrement the count value
  goto  Loop        ; Jump back to Loop if Count != 0
bsf     STATUS, Z  ; Set Zero flag to indicate loop end
```

Immediate arithmetic operators

The immediate arithmetic operators use explicitly specified values for the operations. These values are actually part of the instruction.

"movlw" instruction

The "movlw" instruction is used to load "w" with a constant value. (See Fig. 4-12.)
The movlw instruction does not change any status registers during it's execution.

"addlw" instruction

"addlw" is used to add an immediate value to the contents of the "w" register. (See Fig. 4-13.)

"addlw" changes the Zero, Carry, and Decimal Carry flags in exactly the same way as the "addwf" instruction.

Note, the "addlw" and "sublw" instructions are *not* available in the 16C5x devices.

"sublw" instruction

The "sublw" subtracts the value in "w" from the literal value. This sounds confusing, and it is. (See Fig. 4-14.)

I feel that the way this instruction, like "subwf," would make the most sense is if the literal value was subtracted from the value in "w." However, the PIC architecture does not support this. The best way of thinking of this instruction is to visualize what happens in your mind. It looks like:

```
w = Literal - w
```

and not

```
w = w - Literal
```

which is what I indicated earlier would be the most intuitive.

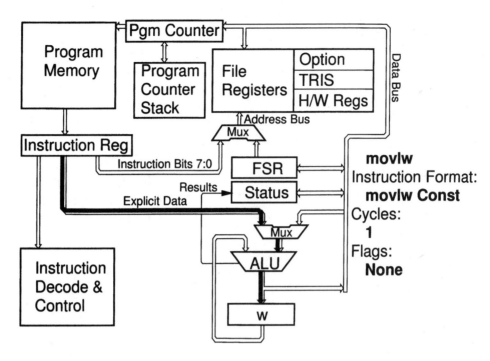

4-12 "movlw" instruction.

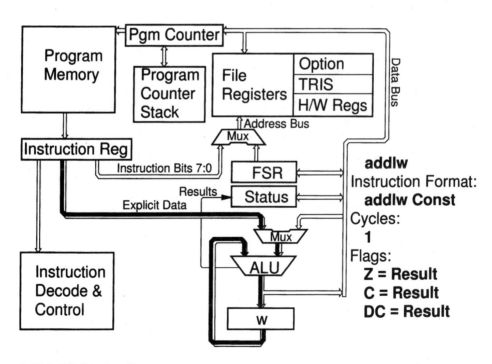

4-13 "addlw" instruction.

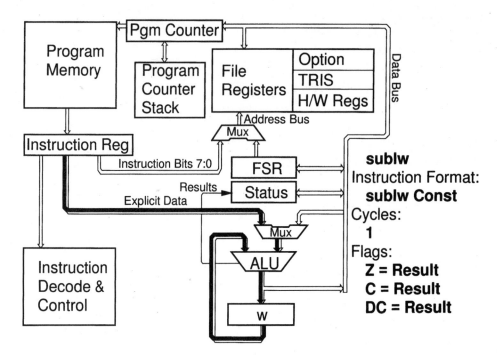

4-14 "sublw" instruction.

Using the derivation used for "subwf," "sublw" is actually:

```
w = Literal + (w ^ 0xOFF) + 1
```

"sublw" changes the flags in a similar manner to that of "subwf."

Because this instruction works in a way that I find to be non-intuitive, I try to avoid using it except to negate the value in "w" by using the instruction "sublw 0."

There is one little trick you might want to use if you have to subtract an explicit value: Add the negative of the value.

For example, say you want to create the code to do the following:

```
w = w - 47
```

This could be:

```
movwf   Temp        ; Save the value in "w"
movlw   47          ; Get the subtraction value

subwf   Temp, w     ; Subtract it from the original "w"
```

If you are using a PIC16C*xx* (not the low-end devices), you can use:

```
addlw   0 - 47      ; Add the Negative Value
```

If you are using a PIC16C5*x* (the low-end, which doesn't have this instruction) and want to negate the contents of "w," an interesting little snippet of code came up on PICLIST one day:

```
addwf   Reg, w      ; "w" = "w" + Reg
subwf   Reg, w      ; "w" = Reg - "w"
```

```
;  "w" = Reg - ("w" + Reg) with above line
;  "w" = Reg - "w" - Reg
;  "w" = -"w"
```

The result in "w" will be the negative of what was in there before. "Reg" will never be changed.

Immediate bitwise instructions: "andlw," "iorlw," and "xorlw"

The bitwise immediate operations ("andlw," "iorlw," and "xorlw") carry out the bitwise logic operations directly on the contents of the "w" register. (See Fig. 4-15.)

These operations, like the register address bitwise operations, only set the Zero flag in the STATUS register depending on the result. The result is stored back in "w," with no opportunity to store the result in a register during the instruction.

Executing the instruction "iorlw 0" is a good way of determining whether or not the value in "w" is equal to zero or not (following the instruction, the Zero flag will be set/reset appropriately).

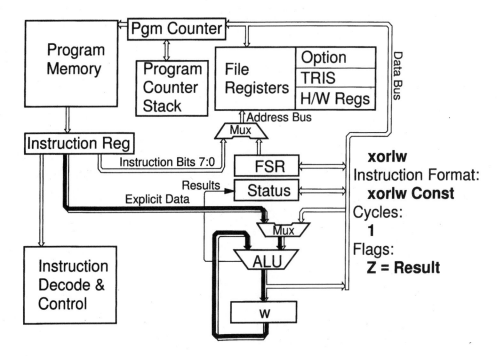

4-15 "xorlw" instruction.

"retlw" instruction

The "retlw" instruction is useful for subroutines that return a condition as well as a table (which is described elsewhere). The instruction loads "w" with an immediate value before executing a return from subroutine. (See Fig. 4-16.)

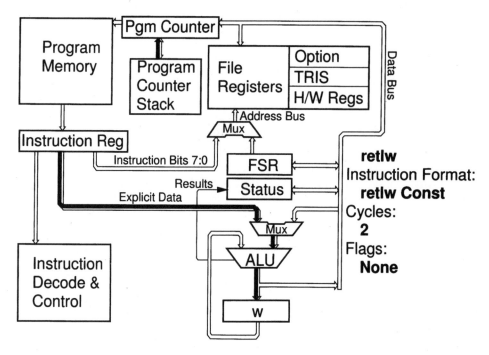

4-16 "retlw" instruction.

The "retlw" instruction can replace the two statements:

```
movlw   Value       ;  Get the value to put in "w"
return
```

The "retlw" instruction is the only subroutine return available to the low-end (16C5*x*) devices.

Execution change operators

Before attempting to use a "goto" or "call" instruction, it is imperative that you understand how they work. If you haven't done so, go back and read about "The Program Counter" in chapter 3. The "goto" and "call" instructions can behave strangely in some circumstances, and the PC will not have the correct destination address.

The reason for the unusual actions taken during the "goto" or "call" instructions is because, in the PIC instruction set, all instructions are the same length. This means that the address size could be larger than what the instruction has space for.

Both "goto" and "call" can be explicitly defined within a device specific "page" (256/512 addresses for the low-end, 2K addresses for the mid-range, and 8K addresses for the high-end). If the label to go to or call is outside the page, the PCLATH register (or appropriate STATUS bits, for the low-end) must be set with the correct page information.

For example, jumping between pages in the mid-range PICs can be accomplished by:

```
movlw   high Label   ;  Interpage "goto"
movwf   PCLATH
goto    Label
```

In this snippet of code, the PCLATH register is updated with the new page before the "goto" instruction is executed. This forces the PC to be loaded with the correct and full "label" address when "goto" is executed.

"call" works almost exactly the same way as "goto," except the pointer to the next instruction is stored on the Program Counter stack. (See Fig. 4-17.)

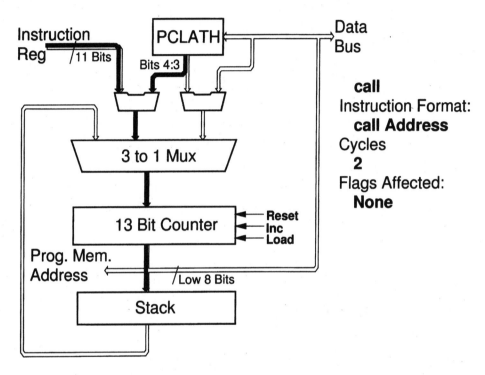

4-17 "call" instruction.

It should also be noted that, in the PIC16C5x family, subroutines to be called can only start in the lower 256 addresses of each 512 address page. This is because the bit definition for the instructions don't allow access to the full 9-bit "page."

There are three different types of return statements in the mid-range and high-end PICs (as noted earlier, the 16C5x can only execute the "retlw" instruction). Each one of these takes the value from the top of the hardware stack and puts it in the Program Counter. These addresses are used to return from subroutines or an interrupt.

The simple "return" statement returns the stack pointer to the address pointed after the call to subroutine. No registers or control bits are changed. (See Fig. 4-18.)

"retfie" is used to return from an interrupt. Actually, it works identically to "return," except that the GIE bit in the interrupt control register is set during the in-

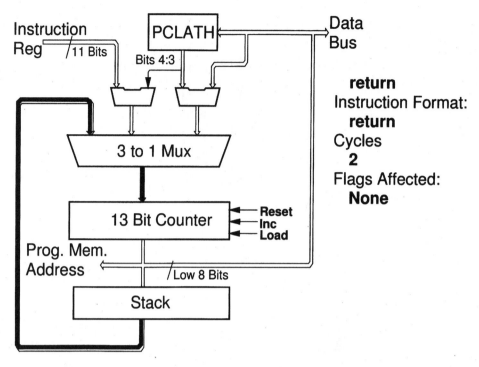

Instruction Reg /11 Bits

PCLATH

Bits 4:3

Data Bus

3 to 1 Mux

13 Bit Counter — Reset / Inc / Load

Prog. Mem. Address /Low 8 Bits

Stack

return
Instruction Format:
return
Cycles
2
Flags Affected:
None

4-18 "return" instruction.

struction. This allows interrupts to occur immediately following the execution of the instruction. (This simplifies the interrupt handler to execute different interrupts sequentially, rather than having to provide a check before ending the handler to make sure nothing is pending and, if there is, to handle them.)

Microcontroller control operators

There are only two instructions that are used to explicitly control the operation of the processor. The first, "clrwdt," is used to reset the watchdog counter. The second, "sleep," is used to hold the PIC in the current state until some condition changes and allows the PIC to continue execution.

"clrwdt" instruction

"clrwdt" (Fig. 4-19) clears the watchdog timer (and the TMR0/WDT prescaler if it is used with the watchdog timer), resetting the interval in which a timeout can occur. The purpose of "clrwdt" is to reset the PIC if execution is running improperly (i.e., caused by an external EMI "upset"). To help ensure "clrwdt" is not executed at an inappropriate time, the PIC should only have one "clrwdt," and this should only be called by executing through one path (i.e., every time an operation has completed and the queue for the next one is about to be checked).

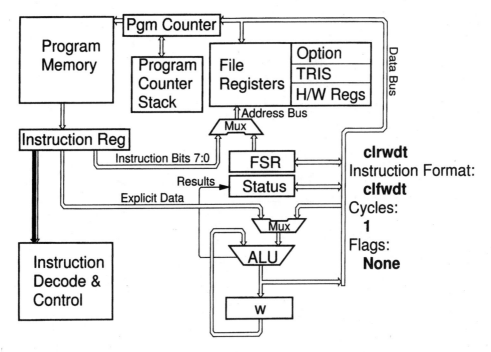

4-19 "clrwdt" instruction.

"sleep" instruction

There are two purposes of the "sleep" instruction. The first is to shut down the PIC once it has finished processing the program. This prevents the PIC from continuing to run and screwing up the other hardware in the application. Using the PIC in this manner presumes that the PIC is only used for a certain aspect of the application's execution (i.e., initialization of the hardware) and will no longer be required in the application. (See Fig. 4-20.)

The second purpose of the "sleep" instruction is to provide a method of allowing the PIC to wait for a certain event to happen. The PIC can be flagged of this event in one of three ways. The first is a reset on the _MCLR pin (which will cause the PIC to begin executing again at address 0). The second is if the watchdog timer wakes up the PIC. The third method is to cause wake-up by some external event (i.e., an interrupt). Using "sleep" for any of these methods will allow you to eliminate the need for wait loops and could simplify your software.

Note that "sleep" takes a relatively long time (1024 clock cycles) to reset the PIC and restart the clock before it will resume executing code. (See Fig. 4-21.)

"option" and "tris" instructions

As noted elsewhere in this book, the PIC16C5x devices only have one register bank (no extra banks) and therefore do not have any way of directly accessing registers that are normally in bank 1 (i.e., the OPTION and TRIS registers). To allow you

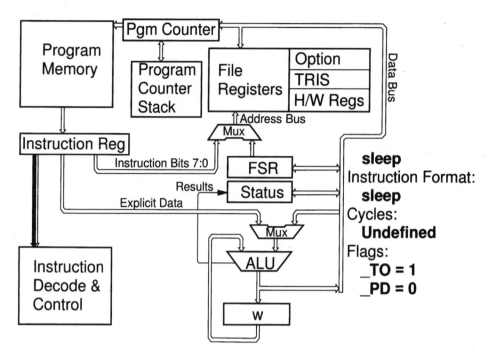

4-20 "sleep" instruction.

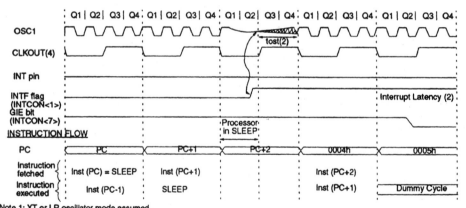

Note 1: XT or LP oscillator mode assumed.
  2: tost = 1024tosc (drawing not to scale). This delay will not be there for RC osc mode.
  3: When GIE = 0 processor jumps to interrupt routine after wake-up. If GIE = 1, execution will continue in line.
  4: CLKOUT is not available in these osc modes, but shown here for timing reference.

4-21 Wake-up from sleep through interrupt.

to access these registers, the PIC instruction set has the "option" and "tris" instructions. Both instructions write the contents of the "w" register directly into the appropriate register. This means that you are unable to read back the contents of these registers. (See Fig. 4-22.)

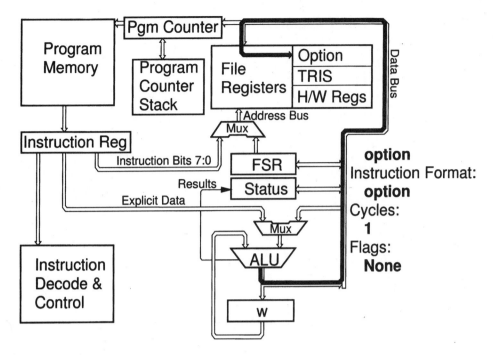

4-22 "option" instruction.

The "option" instruction is quite straightforward, but the "tris" instruction (Fig. 4-23) does merit some discussion. In all PIC microcontrollers, there is more than one I/O port. You're probably wondering how the TRIS registers are accessed. In the "tris" instruction, the port to be used is specified (with a numeric value of 5 to 7 or the port name [i.e., "PORTA"]) in the instruction.

To write the contents of "w" into "PORTB," the instruction would be:

```
tris    PORTB
```

These two instructions are currently available in all the mid-range PIC devices. However, they are only necessary for the (low-end) PIC16C5x devices. Use of these instructions in the mid-range PICs (which have multiple banks) is not recommended by Microchip because the instructions might not be available in future versions of the PIC.

"nop" instruction

"nop" means no operation; when executed, the processor will just skip through it, and nothing will change. Actually, if you study a bunch of different processors, you will find that they all have a "nop" instruction. (See Fig. 4-24.)

I always smile when I hear the term "nop" because of someone who was once called by a co-worker a "nop." When I asked why the co-worker called him that, I was told it was because the person took up space and wasted time. So, even if you don't use "nops" in your software, you've now got a really neat way to insult somebody.

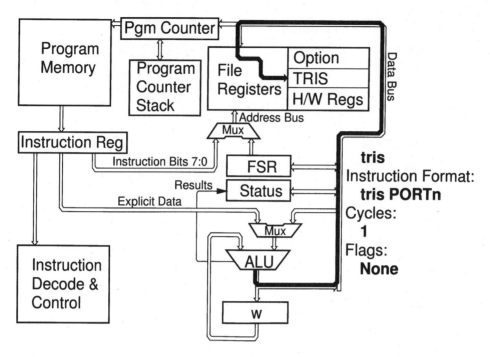

4-23 "tris" instruction.

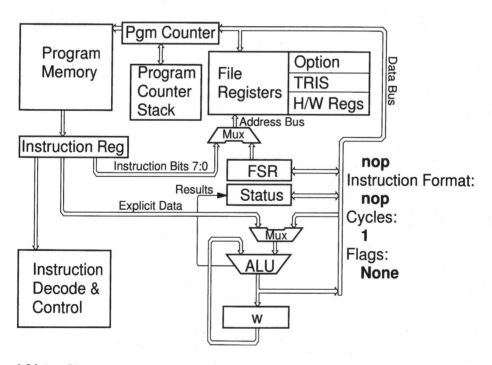

4-24 "nop" instruction.

"nops" traditionally are used for two purposes. The first is to provide time synchronizing code for an application. If you look in "serial.inc," you'll see that I use them to synchronize the output of data out on the serial line (i.e., make sure 1s and 0s happen at exactly the same time within the data output).

The second traditional use of "nops" is to provide space for "patch" code. This is usually done by replacing the "nop" with a patch instruction. In the PIC, it is inconvenient to use "nops" in this manner. This is because of the way the programmable memory used in the PIC works. In EPROM (and EEPROM) technology, when the memory is ready to be programmed (i.e., "cleared"), all the cells are set to one. During programming, the zeros are added to make the various instructions.

This makes it a problem for the PIC because the "nop" instruction is all zeros. This means that an instruction cannot be burned out of a "nop."

However, there is a method of providing space in the code for patches. This means that the code loaded into the PIC can be modified without having to erase the device (which can take a while for an EPROM device) and then reload the code. To do this, the reverse of making instructions from "nops" is used.

For example, you can put the following code in your PIC program to provide patch code space:

```
goto    $ + 5       ;  Skip over five patch addresses
addlw   0x0FF       ;  Instruction with all bits set
addlw   0x0FF
addlw   0x0FF
addlw   0x0FF
```

To enter some patch code, all the 1s in the "goto" statement can be turned to 0s, making the instruction into a "nop." The "addlw 0x0FF" lines are used because they have all the bits set. This construct will allow you to add up to four instructions without having to reassemble your code.

As in the previous example, all mid-range PICs use "addlw 0x0FF" to make sure all the bits are set at a program address. Low-end PICs use "xorlw 0x0FF," and the high-end use "call 0x01FFF" to achieve the same effect.

Because the "nop" instruction, in the PIC, consists of all the bits set to zero, instructions can be programmed out very easily, allowing deleting sections of code without resorting to erasing and reprogramming an entire device (which might not be possible in the case of an OTP part).

Bit operators

Bit operations in the PIC are rather special in that they affect only one bit at a time. There are two types of bit operations. The first is the setting or resetting individual bits. The second is a skip instruction based on whether or not a bit is set.

Bit set/reset instructions

Bit setting and resetting is done by the "bcf" and "bsf" instructions, respectively. These instructions can be thought of in two ways. (See Fig. 4-25.)

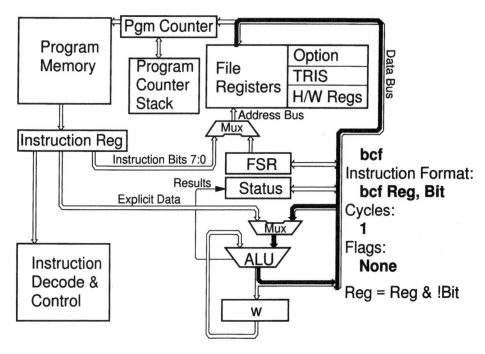

4-25 "bcf" instruction.

In traditional programming, there are times when a value is ANDed or ORed with a constant when only one bit is affected. This can be modelled very easily using the "bcf" and "bsf" instructions.

For example, you want to AND a value with 0x07F. Normally this would be done by:

```
movlw  0x07F      ; Load in the value for ANDing
andwf  Reg, f     ; AND it to get a single value
```

However, in the PIC, this can be simply done by one instruction:

```
bcf  Reg, 7       ; Reset bit 7 of the register
```

ORing works with the "bsf" (bit set) in an analogous way. Any bit in the register Space that is writeable can be handled this way. These instructions can be used to great advantage in situations where you only want to change one bit in a register (such as in the STATUS register).

In using this instruction, not only is the operation carried out in one instruction cycle, rather than two, but the operation can be carried out guaranteeing that no other bits in a register are affected.

Actually, this is a lie. As explained in chapter 3, the use of "bsf" or "bcf" might not be appropriate with I/O ports. This is because of the inadvertent changes that can take place with the register output latches after executing this instruction.

Skip on register bit condition

As noted earlier, the PIC architecture doesn't use jump on condition instructions. Instead, there are a number of instructions that allow skipping the next instruction in line. As noted previously, "decfsz" and "incfsz" can be used for loop control. What I wanted to talk about here is actually controlling the execution of the program.

Program execution control is carried out by use of the two instructions that allow the program to skip the next instruction if a certain register-bit condition is met. In a traditional architecture, jumps and branches are controlled by STATUS register conditions. This can be done in the PIC as well by using the bit skip instructions and accessing the STATUS register.

For example, if you wanted to jump if the Zero flag was set to zero ("jz" in Intel i86 parlance), the following PIC code would be used:

```
btfsc  STATUS, z  ; Skip next instruction if zero bit is reset
goto   label      ; Zero bit was set, goto the specified location
```

Bit skip instructions are useful in a variety of cases, from checking Interrupt Active flag bits to seeing if a number is negative (checking the most significant bit). In chapters 8 and 9, you can see a number of different ways in which the bit skip commands can be used to simplify the software development.

The bit commands are quite unique to the PIC compared to other microcontrollers. I believe they give the application developer a significant advantage in developing applications, allowing far easier methods of bit manipulation that would be otherwise possible.

High-end (17C4x) PIC instructions

If you are using the high end PICs (17C42 and above), you'll probably note that I haven't explained several of the instructions used for these processors in this chapter. (The 17C4x instruction set has 55+ instructions, whereas the low-end and mid-range devices have roughly 30.)

There are three types of enhanced 17C4x instructions: additional arithmetic instructions, data movement instructions, and table read and write instructions. I won't go through the additional arithmetic instructions because they behave in a very similar manner to the mid-range instructions explained in this chapter. They simply provide enhanced function (for example, the 8×8 bit multiplication) to the high-end PICs over the low-end and mid-range devices.

Data movement instructions replace the "movf" instruction in the other PICs. The two instructions "movfp" and "movpf" are used to pass data back and forth between the primary registers and the file register banks.

For example, the "movfp" instruction looks like Fig. 4-26.

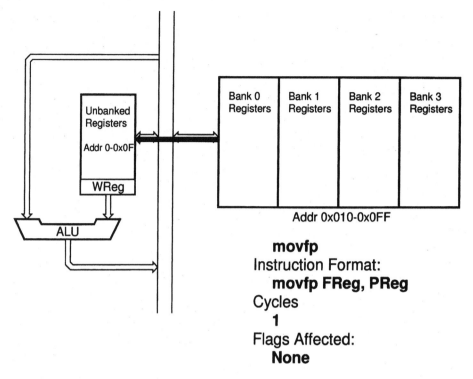

movfp
Instruction Format:
    movfp FReg, PReg
Cycles
    1
Flags Affected:
    None

4-26 "movfp" instruction.

High-end table instructions are really beyond the scope of this book. Simply put, these instructions allow accessing a large table in program memory (essentially blurring the barrier between control store and data space). This program memory typically consists of RAM chips wired to the PIC using the diagram in chapter 3, but it can also be memory-mapped I/O devices, passing back and forth data 8 bits at a time. To do this, the other 8 bits of the 16-bit transfer cannot be active at the same time (to avoid any opportunity for bus contention).

To carry out table reads and writes, first the pointer to the address of the table is set up. This is followed by a 16-bit transfer between the PIC (using the "Table Latch" registers) and the external device.

The "tablwt" (Table Write) instruction works like Fig. 4-27.

The 17C4x, for the most part, behaves like the other PICs, but it does have some features that improve its usability and allow more complex operations.

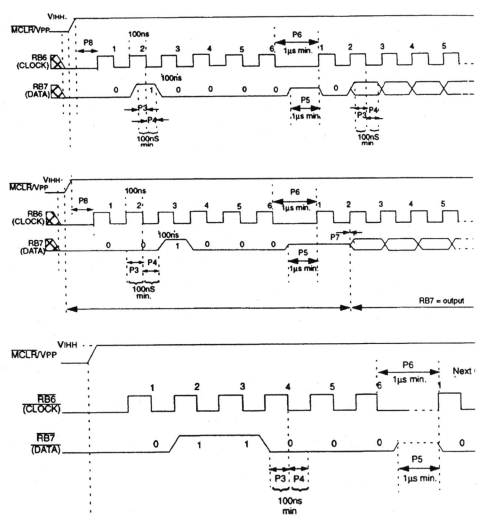

4-27 Mid-Range PIC Programming Waveforms.

5
CHAPTER

Application design

Application design for the PICs has been made a lot simpler by the robustness of the design and the attention to "real world" operating conditions by the PIC developers. This means that developing applications for the PIC is fairly easy (which also means "cheap"). However, there are a few rules that must be followed to ensure a successful and reliable application.

Along with the "few rules," I've also given my feelings on various subjects. Please do not feel that I am the ultimate PIC authority (nothing could be further from the truth). My comments are based on my experiences and understanding of the PIC microcontrollers.

Power input

One of the nice things about the PIC microcontrollers is the wide range of input voltages the devices can accommodate. In my own experimenting, I have found that PICs can handle a much wider range than they are rated at. I have run a PIC 16C84 rated at 4.0V to 6.0V at 2.5V comfortably). However, designing a product to run a 5V PIC at less than 5V for Vdd is definitely *not recommended*.

If you want to develop an application that uses battery (i.e., 2.5 to 3 Volts) power, then there are PICs designed for this voltage, or you can use a step-up power converter to provide 5V to the PIC.

This doesn't mean that power can be taken for granted. The PIC requires reasonable decoupling (0.1 μF) between Vdd and Ground. This is because fairly large transients can be generated during output switching of the PIC. In chapter 8, one of the experiments will be running without a decoupling cap to show how switching transients can affect the PIC's operation.

System oscillators and clocks

Along with a great deal of flexibility on power, the PICs can also use a number of different clocking schemes. These allow a great amount of flexibility in your application design and allow you to meet any specified requirements.

There are four different clocking modes.

The first is using the charge/discharge cycle of an resistor/capacitor (RC) network. This type of oscillator/clock is best used for low-cost applications where the actual timing is not critical (you can reasonably expect to be within 20% of the expected value with an RC oscillator). The target frequency is specified using the figures contained within the PIC datasheets. (See Fig. 5-1.)

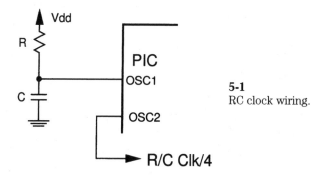

5-1
RC clock wiring.

(*Note:* To ensure stable operation of the RC oscillator, the resistance value should be in the range 3K > R > 100K.)

You will notice that I do not give a formula to allow you to specify the resistor and capacitor values for a given frequency. This is on purpose because a generalizing formula (like f = 2.2/RC) does not apply for the PIC because the charge/discharge path does not consist of only linear components. Instead, you should look at the datasheets for the PIC used in the application and use the recommended resistance and capacitance values for a given frequency from that.

The RC oscillator is best for applications where the timing is not critical and cost is. If you wanted to use an RC oscillator (and the reasons for doing so would probably be focused on the robustness of the oscillator compared to some of the other solutions), you could tune it for exactly the frequency you required. I would like to discourage this type of plan because the end result will not be very reproducible (imagine a modern, automated factory with dozens of people trying to match caps to resistors to PICs). If your application is going to see any kind of volume (say greater than one) and you require an accurate clock, you should look at the solutions that follow.

The next step up from RC oscillators are *ceramic resonators*. These are simple devices that give you much better accuracy than the RC (typically within 0.5% of the expected value as opposed to the RC circuit's 20% to 30%). This accuracy will be good enough for most tasks that you require (including serial communications). The ceramic resonator has excellent mechanical durability, which makes it appropriate for mobile applications (such as the "I/R tank" in chapter 9).

The only designs I would say a ceramic resonator would be inappropriate in is in applications where a very accurate time base is required (i.e., a clock or something that plays music).

The clock that you are using must be identified before the PIC is programmed so that the configuration fuses can be set. The RC oscillator uses its own specific configuration, while the other types of oscillators can use any of the three different specifications; the differentiator being the speed that is used.

Ceramic resonators are easy to set up and have a low part count. (See Fig. 5-2.)

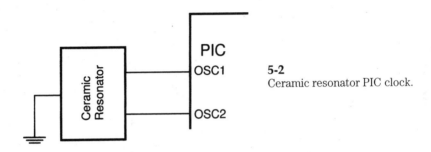

5-2
Ceramic resonator PIC clock.

Crystal oscillators provide the best in timing accuracy. These devices are still quite cheap and will give you outstanding accuracy. They are, however, a bit more finicky to set up. (See Fig. 5-3.)

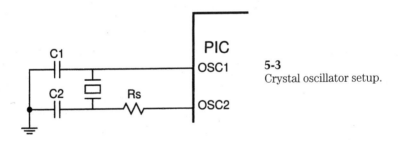

5-3
Crystal oscillator setup.

Extreme care must be taken in using a crystal to ensure that the load capacitance for the circuit is correct. Figure 5-4, which was taken from the PIC databooks, will give you an idea of what cap values should be used for different frequencies.

As noted elsewhere, I have not had an application requiring Rs. If the PIC's oscillator doesn't start up reliably, you might want to experiment with an Rs value of 100 to 200 Ω.

Mode	Freq	OSC1	OSC2
LP	32 KHz	33–68 pF	33–68 pF
	200 KHz	15–47 pF	15–47 pF
XT	100 KHz	47–100 pF	47–100 pF
	500 KHz	20–68 pF	20–68 pF
	1 MHz	15–68 pF	15–68 pF
	2 MHz	15–47 pF	15–47 pF
	4 MHz	15–33 pF	15–33 pF
HS	8 MHz	15–47 pF	15–47 pF
	20 MHz	15–47 pF	15–47 pF

5-4 Capacitor Selection for Crystal Oscillator for PIC16C71

As a rule of thumb, with the PIC, using less capacitance will give you a better waveform. If the capacitance is too large, your clock will degrade to the point where the PIC won't work. If you aren't sure about the capacitance values you are using, take a look at the clock waveforms at the "OSC2" pin using a high-impedance oscilloscope probe. You should see a waveform similar to Fig. 5-5.

5-5 Good OSC2 waveform.

If you have overloaded your circuit with too much capacitance, you will see something like Fig. 5-6, if the PIC runs at all. (If it doesn't, the signal will be a dc voltage at approximately Vdd/2.)

5-6 Bad OSC2 waveform.

You might find that crystals are not very robust; they don't take well to being knocked around. For this reason, you might want to try a ceramic resonator once your application is running with the crystal.

The last type of clock that can be used for the PIC is a TTL clock. This can be provided from a specific oscillator "can" or from a convenient clock found in your circuit. In any case, only the OSC1 pin is used for hooking up the oscillator. This leaves OSC2 floating or unconnected (see Fig. 5-4).

Reset

Like power and clocking, reset can be very simple as well, as long as a few rules are observed. The _MCLR line should be held low until power can be reasonably assumed to be good. There are a number of circuits that are good for doing this.

It is important to understand where the PICs begin to execute at power up and other reset conditions (i.e., _MCLR reset, wake-up from "sleep," and a watchdog timer timeout). In the low-end PICs, the first address executed is the last address in PIC's address space (i.e., for the 16C54, the address is 0x01FF). This means that the first instruction the PIC will execute is always a jump (to the power-up code). The instruction at this address can be set to a "nop," and the Program Counter will roll over and start executing at address 0.

In mid-range and high-end PICs, the reset address is always zero. This means that the code begins to execute at the beginning of the address space and goes from there. You might not be able to start putting your power-up code at this location if you are using interrupts. Interrupt vectors start at address 0x04, which means that the power-up code must be jump around this address.

Using the built-in Power Up Timer might actually give you a chance to use no reset circuit at all! (See Fig. 5-7.)

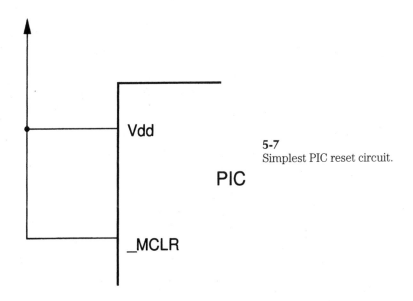

5-7
Simplest PIC reset circuit.

The Power Up Timer holds off the PIC Reset for 72 milliseconds to allow power to stabilize. If you have an application with a very stable power supply (i.e., battery powered), you might want to forgo using any type of reset circuit at all and just hold Reset (_MCLR) to Vcc and let the PIC wait until the power is stable.

While the circuit in Fig. 5-7 is the simplest, I highly recommend putting a resistor pull-up between Vdd and _MCRL. (See Fig. 5-8.)

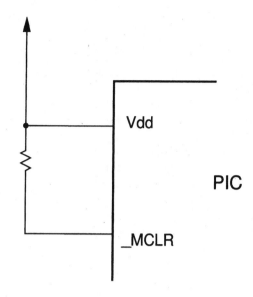

5-8 Simple reset with pull-up.

One of the side benefits of this circuit is that the 18-pin PICs have _MCLR besides Gnd. This means that, during debugging, when you want to reset the PIC to observe how it powers up and operates, you can simply short out _MCLR to Ground (my tool of choice is a small screwdriver).

The power-up sequence looks like Fig. 5-9.

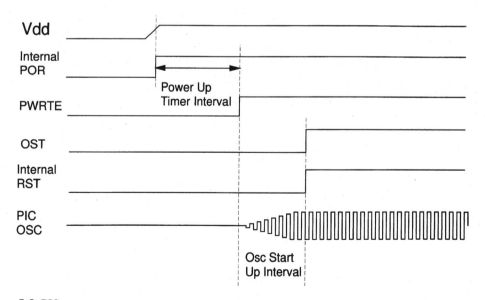

5-9 PIC power-up sequence.

Note that, if this reset circuit in Fig. 5-8 is used, the Power Up Reset Timer (which causes a 72 msec delay between power up and the PIC running) bit in the configuration fuses should be set.

For critical applications or applications where the power supply cannot be guaranteed (i.e., a battery is used to power the circuit), other reset circuits should be used. These are generally known as "brown-out" protection circuits.

These circuits generally focus on providing a comparison between Vdd and the minimum operating voltage the PIC requires. If Vdd falls below this point, the PIC is reset to prevent erratic operation. If the PIC will be operating in an environment where this is possible, you should make sure the code supports repeated restarts without causing problem with the peripheral hardware.

In the PIC datasheets, there are a number of different examples of active reset circuits to reset the PIC when the voltage is too low to reliably run the PIC. Along these examples are a number of simple devices which can be used for this purpose. (See Fig. 5-10.)

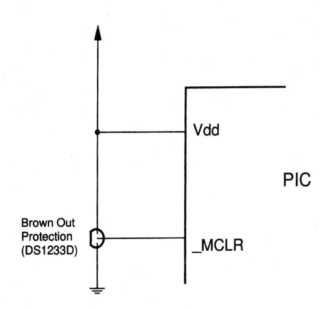

5-10 Reset with brown-out protection.

The advantage of these devices is their small footprint and simplicity in adding them to a circuit. I should point out that some of the PICs do have brown-out protection built in (making this circuit redundant). Enabling this hardware will allow you to simply tie _MCLR to Vdd.

The _MCLR pin can also be driven by a standard TTL or CMOS driver for external control over the operation of the PIC.

Interfacing to external devices

In this section, I will discuss some basic types of interfacing and the rules that go along with them. The material will be at a very general level with "rules of thumb". Applications can be found later in this book.

TTL/CMOS

Obviously, interfacing to digital devices will be the simplest way of hooking up a PIC. The most basic way of doing this is moving the data in and out in a parallel or serial manner using the hardware provided in the PIC.

This can be expanded for communicating directly with more than one peripheral device as required. There are a few rules for doing this.

Determining the type of bus is the first priority. A simple input or output bus can be accomplished easily. Implementing a bidirectional bus requires some arbitration rules for both input and output devices.

A parallel output bus is most easily implemented using device selects on the bussed devices to address the device you want to talk to. This is similar to a "typical" microprocessor application after the addresses have been decoded.

This method of controlling parallel devices is used in the blinking Christmas tree lights project in chapter 9.

Input from parallel devices can be done two different ways. The first is to use open collector drivers or switches pulling down to ground a pulled-up bus line. This is often known as a "Dotted AND" bus (because when one or more devices makes the line active, or low, the entire bus is low). Using this type of bus does offer one significant advantage: Multiple devices on one line can output data to the PIC simultaneously. This makes this type of bus particularly attractive for external interrupt sensing.

Some of the PIC's I/O pins are an "open collector" (actually "open drain") drive for this purpose. Any other type of logic drivers will cause problems driving the line in parallel causing *bus contention*, resulting in an unknown logic state. Along with the open collector drivers, a pull up to Vdd is required (this is available on some PIC I/O Pins). (See Fig. 5-11.)

The second method of implementing the input bus is to use devices that have output that can be put in a high-impedance state. This means that, when the PIC is ready to read from a particular device, it must tell the device that it is ready for the device to drive the bus. The pin normally used on the peripheral device is called a "read" pin and is usually negatively active.

This method of inputting data is actually a simple processor bus, as is shown in Fig. 5-12 (with the PIC operating as an intelligent peripheral).

In implementing this type of input bus, there are a few things to watch out for. The first is that you must make sure that all the devices on the bus can have their outputs put into a high-impedance (high-Z) state. Any output drivers that can't go into a high-impedance state can result in bus contention. The other important thing

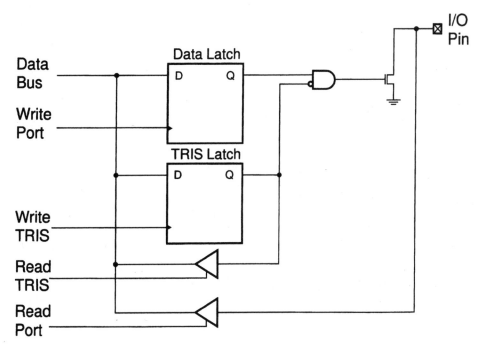

5-11 Common collector pin diagram.

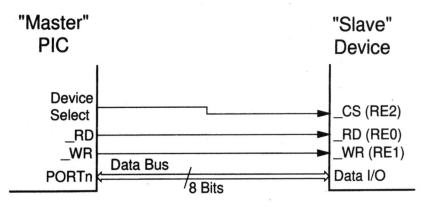

5-12 Parallel connection to external devices.

to check for is to make sure that all your devices can drive data for the length of time required by the PIC (and your software) to read it. This is important because some devices can only output data for a microsecond or two.

The code required to read a bus device would be:

```
bcf      CTL_PORT, _Dev_Read   ; Make the read bit low
movf     DEV_PORT, w           ; Read the device port
bsf      CTL_PORT, _Dev_Read   ; Turn off device output drivers
```

Creating a read/write (bidirectional) bus will involve using the instructions for the bus write and bus read (with high-Z drivers) with a few extra wrinkles. In the two unidirectional bus modes, the chip selects were used with the read/write pins on the peripheral devices. For a bidirectional bus, the chip select bits (usually negatively active) should be controlled independently of the data bits and the read/write control bits. This is shown in Fig. 5-12.

I always return the PIC output port to "read" (i.e., all the TRIS bits set) after completing the write. This will prevent any chance of bus contention due to the PIC driving the bus at the same time as one of the other peripheral devices.

Now, having said all this, you might want to skip all this and use a high-end PIC (17C4*x*) in your application. The high-end series of PICs have an Extended Microcontroller Mode in which peripheral devices could be put directly on an external PIC bus. The Extended Microcontroller Mode uses a multiplexed address/data bus that complicates the design slightly. (See Fig. 5-13.)

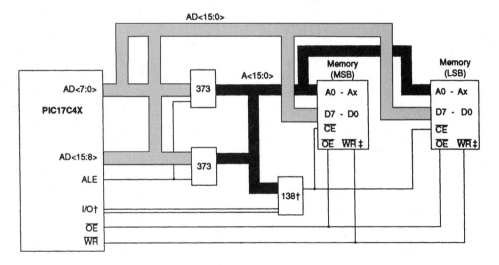

† Use of I/O pins is only required for paged memory.
‡ This signal is unused for ROM and EPROM devices.

5-13 Typical external program memory connection diagram.

This is a slightly different application than what is shown in the databooks. Putting peripheral devices on a processor's data bus along with memory devices is known as *memory-mapped I/O*.

After reading all this about busses, you must think that interfacing to other devices is simple. For the most part, you're right. There are, however, a few things for you to watch out for.

In the applications, you will see that I generally use standard TTL parts (usually 'LS244s and 'LS374s) for input/output. These devices use the PIC I/O pins for data transfer and control. These parts are easy to obtain and are generally very cheap. Using these devices, the I/O can be expanded to Parallel to Synchronous Serial Data Input and Output. This is shown in chapter 8.

Analog input/output

There are two different types of ADCs used on the PIC. The $16C62x$ family uses a voltage comparator circuit, while the $16C7x$ devices use a timed capacitor charging circuit. In terms of electrical input characteristics, the two methods of onboard analog-to-digital conversion are quite similar. The differences are most evident in the applications the circuits are best suited for and the software used to drive them.

There are two things that are critical to be aware of when using the $16C7x$ for analog voltage measurements. The first is to make sure you allow enough time for the voltage to be loaded into the PIC's internal storage capacitor and allow the ADC to determine what the input voltage actually is. This process takes about 40 msecs. The voltage source to be measured shouldn't have an output impedance greater than 10K to allow the internal cap circuit to charge properly (and in a reasonable amount of time).

The $16C62x$ devices contain an onboard analog voltage reference that can either be output from the device or used with an analog comparator to look at an input voltage and determine whether or not the input is greater to or less than the reference voltage. This circuit allows very fast comparison of input voltages (i.e., for a thermostat) and notification of when the voltage exceeds the set limits.

The $16C62x$ devices can be used to determine an input level, although it would require software to develop an approximating routine. If you have an input voltage that can be of radically different values, the $16C7x$ family is much better suited for measuring the value.

The $16C62x$ can also be used to generate analog output voltages to control other analog devices. This is done by primarily two different methods. The analog voltage signal is produced from the PIC using an digital-to-analog converter (DAC). This device selects the correct voltage, using a multiplexer, from a number of resistor-based voltage dividers. This voltage can either be used inside the PIC as a reference voltage (which is why it is described as "VRef" in the datasheets) or outside as an analog voltage output. (See Fig. 5-14.) An analogous method of analog output is also shown in chapter 8.

The other method is by Pulse Width Modulation (PWM) in which a digital string is output from the device as shown in Fig. 5-15.

This series of pulses is used to control the speed of a dc motor or the position of a servo. PWM signals (known as *pulse trains*) can also be created internally in the

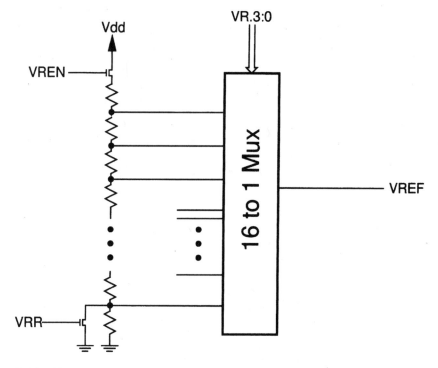

5-14 16C62*x* VRef circuit.

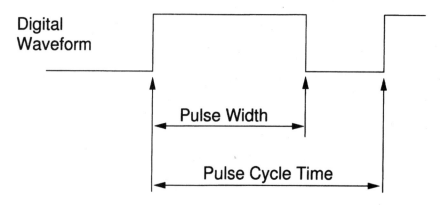

5-15 PWM pulse.

PICs that have TMR2 (which is specially designed for creating PWM signals for external devices).

Examples of 16C7*x* analog input and PWM output are shown later in the book.

Different logic levels

Often in applications, you will have to translate data between different logic levels. The classic example and the one you will probably encounter the most is interfacing PICs to other devices (i.e., PC's using RS-232). In your career, you will probably have to interconnect a number of different logic families and, while the science behind this could fill an encyclopedia, I just wanted to touch on a few important points.

RS-232

There are a number of level translators for RS-232 to TTL/CMOS. These devices range from the "original" 1488/1489 Buffers (which require +/–12V to translate the signals properly) to single 5V supply components (some of which are used in projects described in this book). These devices ensure that the RS-232 voltages are valid and do not violate any specifications. Personally, I feel this is the best way to interface TTL/CMOS logic to RS-232 levels.

Depending on the application, you might want to use an even simpler method of interfacing to an RS-232 source. This method involves simply hooking the PIC up to the RS-232 output of another device using a simple series-terminating resistor. Actually, it's not that simple, and before attempting this, a clear understanding of what is being done is necessary.

Figure 5-16 shows the levels of RS-232 and the PIC CMOS.

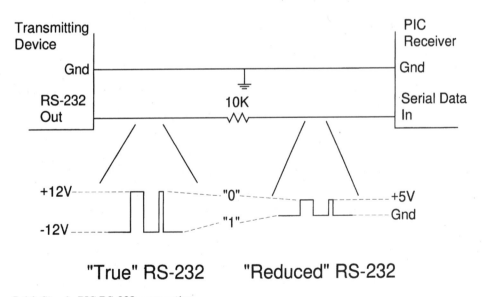

5-16 Simple PIC RS-232 connection.

From Fig. 5-16, there are two problems that must be solved before the PIC can be connected to the RS-232 line. The first is the polarity. By looking at the RS-232

Specification, you will see that a 1 is a negative voltage, while a 0 is a positive one. This contrasts directly with the closest PIC logic levels. A 1 is typically defined as a positive voltage, and a 0 is defined as a zero voltage. To rectify this problem, the receiving software can invert the bit levels and treat the values accordingly.

Because of this logic inversion, this means that the built-in asynchronous serial receivers/transmitters available in some PICs cannot be used (because they expect the data to come in a positive format). This also means that this method of handling RS-232 is really restricted to using I/O ports that do not have the asynchronous serial hardware.

The second problem is the difference in the logic levels. Anytime a signal is sent that is greater than or less than what is expected by the receiving logic, there is the opportunity that a damaging current flow will occur. This can be prevented by putting in a current-limiting resistor (I typically use 10K) in series as shown in Fig. 5-16.

This approach works very well for receiving data from an RS-232 transmitter. Using the reverse to drive a signal onto an RS-232 line is not quite so simple. A valid "0" is between 3V and 15V. The most the PIC can put out is 5V or so, which, while a valid signal is very close to the "indeterminate" range and might not be read properly by the receiver. Conversely, a 1 is a voltage between –3V and –15V. Some RS-232 receivers might not read the PIC output of zero volts as a valid 1.

For this reason, I do not recommend that this method be used for PIC RS-232 transmission. While many different RS-232 serial ports are able to handle these levels, many do not (including the PC that I used to create the experiments and projects for this book on). In chapter 9, I show two different commercially available parts that will allow simple RS-232 level conversion.

ECL and level shifting

As noted previously, there will be times that you have to interface a PIC to different logic levels. If you reread the previous section on receiving RS-232 signals through a single current limiting resistor, you will get a hint of what I am going to suggest here.

There are interface chips for interfacing different logic families; however, these chips tend to be expensive and difficult to procure. Instead, the most common way of interfacing families is by use of multiple power supplies and *shifting* the level of the logic to match the receiver.

This is done by changing the power of one of the devices as in Fig. 5-17.

As you look at Fig. 5-17, something should become clear: As the PIC's power supply ground reference is shifted negatively, the *threshold* (or switching) level of the device is being shifted to be equal to the ECL's switching threshold. The switching level is the voltage where the input value read in changes value (from a 1 to a 0 or vise versa). By matching up the switching levels, different logic swings can be accommodated easily. The current limiting resistor in series is to prevent large current flows between the different logic families.

Shifting is not required for RS-232 because the switching level of the PIC (at 1.5V) is inside the switching voltage range of RS-232. (RS-232 has a switching region of –3V to +3V in which no value (0 or 1) is valid.)

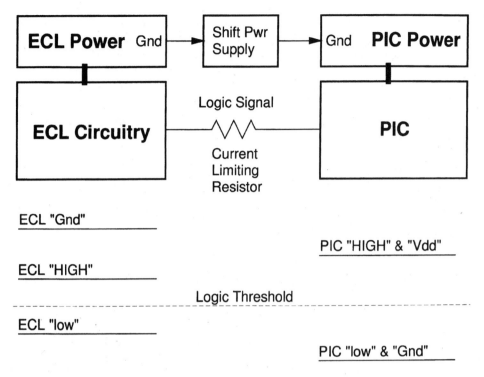

5-17 ECL to PIC level conversion.

Care must be taken to ensure that the power supply level shift is reliable. This is usually done by placing a low-resistance load (1 to 10 Ω) across the output of the "Shift Power" supply. The low-resistance load will ensure that the ground power supply will operate and regulate properly, giving a stable Ground reference for the shifted device. Make sure that the low-resistance load can handle the power that will be going through it.

Interfacing devices of different logic families is not something that is recommended for applications. This is because the design might require some tweaking to get reliable operation for a given set of hardware. Instead, this is best used in "proof of concept" prototypes and applications that you can control. However, I have done this in manufacturing test equipment where developing custom circuits or buying level converters was not appropriate in terms of cost or scheduling.

Interrupts

In applications that I have seen, interrupts are generally one of the least used features of the PIC microcontrollers. This is a mystery to me because they are quite easy to implement and program. I like to think of interrupts as a way of multitasking the processor; you have a mainline and have code that only executes when it is appropriate.

What is an interrupt? Imagine you are engrossed in watching the "X-Files" on TV and the phone rings. What happens? First off, your concentration is broken and you decide what to do. Are you expecting the call? If you are, then you mute the TV, start the VCR, take the call and watch the taped show when the call is finished. If you aren't and choose to ignore the call, then you let your answering machine take the message.

The big point is that you were interrupted. You handled it and didn't loose the main thread (i.e., finding out what this week's monster is going to do to Mulder and Scully).

PIC interrupts (and, actually, all interrupts for microprocessors and micro-controllers) are very similar. Once the interrupt occurs, everything else stops, and the focus is put in the interrupting condition. The interrupt logic decides how to handle the interrupt, process it or ignore it, and then return to what it was originally doing.

The interrupt "vector" address for the mid-range PIC is 0x04. The *vector* is the address the PIC changes its address to when the interrupt occurs. This means that interrupts from multiple sources must be handled separately in software. An advantage of the PIC hardware is that interrupts are never lost, so if you have two interrupts occurring at the same time, your software can handle the incoming one with the highest priority and return from the interrupt and the interrupt handler will be immediately re-entered with only the lower priority interrupt pending. This eliminates the requirement for storing concurrent interrupt sources to prevent losing them (something that is a problem with other devices—most notably the IBM PC's interrupt hardware).

The interrupt timing looks like Fig. 5-18.

The high-end devices have four different interrupt vectors. This means that the software effort of prioritizing incoming interrupts is greatly reduced (but not eliminated because each vector has more than one interrupt source). In the high-end devices, care must be taken to handle concurrent interrupts and make sure the RAM registers used to store the incoming status values are not overwritten (if the same ones were used for each of the interrupt types).

I've always felt that how you set up and return from an interrupt is at least as important as how you handle it in the first place. In chapters 8 and 9, I will demonstrate interrupt handlers. However, I wanted to talk about the code used for initially handling the interrupt as well as returning.

Entry into the interrupt handler is important because you want to save everything you were doing exactly. Status information that is lost cannot be re-created. Compounding this is the PIC's lack of a data stack.

The code that follows will save all the important information during the start of an interrupt. The critical values are, what's in the "w" and STATUS registers and any other registers that might be changed within the interrupt handler. As well, it is important to understand what kind of conditions are present at the start of the interrupt.

For this reason, I always use the same interrupt handler header and footer. The following code is not something that is earth shaking; you'll probably see it used in pretty well every application that uses interrupts. The purpose of the header and the

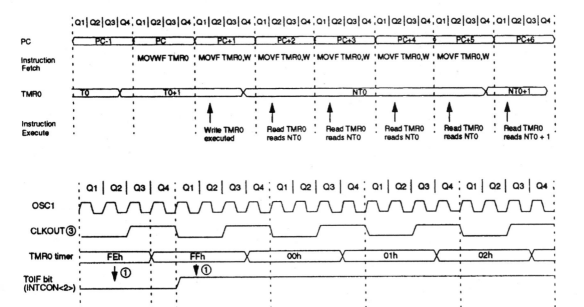

5-18 TMR0 interrupt timing.

footer is to make sure that the "w" and STATUS registers are in exactly the same state as when the mainline program was interrupted.

Here's the code for the interrupt handler header:

```
Interrupt:
    movwf   _w              ; Save the contents of the "w" register
    movf    STATUS, w       ; Get and save the contents of the STATUS
    bcf     STATUS, RP0     ; Optional - Put PIC regs into known page
    movwf   _status         ; Save STATUS register for int return
    movf    PCLATH, w       ; Optional - Save PCLATH
    movwf   _pclath
    movlw   HIGH Interrupt  ; Optional - Set the correct PCLATH for
    movwf   PCLATH          ; the interrupt handler
    movwf   FSR, w          ; Optional - Save INDEX register
    movwf   _fsr
```

Looking through the setup, the first line should make sense to you. I'm just saving the contents of the "w" register. The next two lines might be a bit confusing, but they actually make a lot of sense (especially when we look at the code before returning from the interrupt).

There is one very important thing to understand and plan for with this, and this has to do with the different processor banks. The interrupt can occur when the processor is accessing registers in any of the banks. This means that, when the interrupt occurs, the "RP" bits in STATUS are at an unknown state. Checking/saving/updating the state cannot be done without either changing the processor status bits or the "w" register. To eliminate this requirement, you must make sure space is saved in each RAM register bank for "_w." In some PICs, such as the 16F84, the RAM registers in bank 0 are "shadowed" (can be accessed from either bank 0 or bank 1) regardless of the value of the RP bits, so this requirement might be irrelevant.

After the contents of the STATUS register are saved (in "w"), the bank select ("RP") bits are set to a known state.

The last six instructions of the interrupt handler are optional according to the device that you and the software use. (Note that this is not required if the program does not leave page 0.) If you are using a device and code that execute in a number of different address pages, then you will have to save PCLATH and reset for the location of the interrupt handler's code. This is the same for FSR if it is used as a variable stack in both the mainline and interrupt handler.

Before returning from the interrupt, it is important to reset the current interrupt source (individual IF bit) and set up what kind of interrupts you are going to accept later. Actually, there are a number of instances where the interrupt handler can be used as a "state machine" to simplify your programming and make it more efficient (the infra-red receiver outlined in chapter 9 will give you an example of that).

After the interrupt is handled, the execution state has to be returned to what it was when the interrupt was first received. This is done by the following code:

```
Int_Return:
  movf    _fsr, w     ; Optional: Restore FSR
  movwf   FSR
  movf    _pclath, w  ; Optional: Restore PCLATH
  movwf   PCLATH
  movf    _status, w  ; Get and store contents of the STATUS
  movwf   STATUS      ; Register from before the interrupt
  swapf   _w          ; Flip and load the "w" register without
  swapf   _w, w       ; affecting the STATUS register
  retfie
```

It's important to make sure that the context registers stay the same regardless of what is done after they are loaded. In the previous code, you'll see that I loaded in the STATUS register and then the "w" register.

If you look through the list of register data movement, you will see that only two instructions allow data movement without affecting status register bits. The first is the "movwf" instruction, which is moving the contents of the "w" register into the specified register. The second instruction is the "swapf" command, which exchanges the most significant 4 bits (nybble) in the specified register with the least significant nybble. In doing this, you have a method of transferring data without changing the STATUS Zero flag.

This code will allow you to execute the interrupt handler almost as easily as if you are running in the mainline. Looking over the previous code, you will note that the interrupt handler will always take the same number of cycles, regardless of what the registers contain. This is an important point in real-time applications where critical timings might be required. If this is the case, then you should make sure that the code used to handle the interrupt always runs the same way and that any conditional branching is checked to ensure that all paths take the same number of cycles to execute.

What I haven't talked about is nesting the interrupts. When an interrupt happens, the GIE (Global Interrupt Enable) bit is reset. The bit is set back on (allowing interrupts to happen) after the "retfie" (Return from Interrupt) instruction is executed. It is possible to turn on the GIE interrupt during the execution of the interrupt handler. *Extreme* care must be taken in making sure that the current interrupt status is *not* lost. The previous code (both for interrupt entry and exit) will not be acceptable for this task. My recommendation in this case would be to use the index (FSR register) like a stack and save the important registers in a stacklike fashion (using a different place for every interrupt entry).

Actually, my real recommendation would be to redesign your application to not require nested interrupts. The PIC is not an ideal architecture for doing this, and if it is required, you would probably be best off looking at other microcontrollers.

Interrupts, while being very useful programming constructs, can make your life much more difficult when you have time-critical code in your mainline (especially if you use a timer time-out to provide an RTC feature). If this is the case, you should disable interrupts before starting the time-critical code and re-enabling them as soon as it has been completed:

```
bcf     INTCON, GIE ; Disable interrupts during time-critical code
  .                 ; Minimum code required for a function
  :
bsf     INTCON, GIE ; Enable interrupts
```

After interrupts are re-enabled, you shouldn't be surprised that the next instruction following the "bsf INTCON, GIE" is the interrupt handler. Interrupts will stay pending until the GIE bit is set, allowing them to execute. In a number of the applications shown in this book, having to turn off interrupts occasionally is often required to make sure that mainline code runs properly.

Please note that this code is not required inside the interrupt handler itself. In fact, if the GIE bit is enabled inside the interrupt handler, there might be a chance that you will have a nested interrupt (i.e., an interrupt will begin processing while another is already executing), which would cause the context saving registers to be lost.

To ensure this doesn't happen, the only instruction that should set GIE in the interrupt handler code is the "retfie" instruction.

6
CHAPTER

PIC application software development tools and techniques

Developing the software for PIC (or any other microcontroller) is a unique experience for most programmers. This is because the application is *all* the software in the system: all memory management, all input/output, all resources, and the complete operating system. This can be overwhelming to new programmers.

In chapter 8, I will give you a lot of the tools for doing this work; however, but to do this, you must understand how to program the PIC.

In chapter 4, I discussed all the instructions, how they work, and what they affect. Now I want to spend some time discussing how you program the PIC and give a few tips that have worked for me in making sure my code is understandable, fast to write, and minimizes the potential for (intermittent and otherwise) problems.

This chapter is devoted in helping you survive the experience of developing an application for the PIC.

Software development tools

What a mouthful this chapter's title is. I originally called it "Programming Tools" but changed it when I realized just how many areas there are to cover. As you will see, I'm pretty laid back about how I think people should develop code. The reason for this (and I'll re-emphasize it over and over) is that I believe that everybody should be allowed to develop code and applications in a manner in which they are most comfortable. This doesn't mean that you should stick with what you know because you like it, but you should find what works for you then exploit and improve upon it.

In terms of the language to be used, this is really up to the programmer. In this book, I use Microchip's MPASM Assembler exclusively. The primary reason for this is because of the price (free) and the fact the Microchip Assembler uses the instructions documented in the Microchip databooks. There are some shortcuts provided in the assembler, and while I touch on some of them, I have tried to avoid them wherever possible to create code and examples that are as easily understandable as possible. I've avoided these shortcuts because they tend to make the code harder to read for the beginner (who is trying to relate back to previous information).

This does not mean that I advocate writing all applications in Assembler—nothing could be further from the truth. There are a lot of good development languages out there, and when you go out on your own and develop code, you should use the language you are most comfortable with.

The language you choose should be able to produce efficient code; efficiency being measured in number of Assembler instructions produced for each line of source code. It should also be able to access all the features and registers of the microcontroller that the application requires. Lastly, the language should be something you are comfortable with; learning a new language for a new processor/system should be a choice, not a requirement. Later in this chapter, I discuss high-level languages and their requirements.

One note on tools: Some vendors supply tools that do not use the instruction format specified in the Microchip databooks. The best example of this is Parallax's Programming Tools, which creates instructions that simulate Intel 8051 instructions. This might be something that makes you feel more comfortable (and using tools that make you comfortable is what this is all about), but before buying the tools, make sure that you understand what you are getting and what the implications are. PICLIST, STAMPLIST, and sci.electronics.design are excellent Internet resources for beginning your research on what is the best language for you.

One tool I have found to be invaluable for developing code on a microcontroller is the simulator. A simulator is a tool that allows the application software developer to try out the code and monitor it during execution with a variety of stimulus.

I always try to get 90% confidence in my code before I try it out on the microcontroller. The only way I can do this is by running the code on a simulator. In this book, I will use the Microchip Simulator in either the DOS command-line format (MPSIM) or in the Microchip Windows Development System (MP-SIM). Along with using the simulator, I will also provide the configuration (MPSIM.INI) file and any required stimulus files.

The purpose of this book is not to explain all the details of the Microchip Assembler and Simulator; Microchip does publishes datasheets and manuals for your use, available in either hardcopy or softcopy (Adobe ".pdf") formats.

Creating and supporting your code

A tool often overlooked by beginners is the editor used to develop the code. I feel the editor you end up using should be totally your choice. I've seen people develop large, efficient, and elegant code and programs using every type of editor you can imagine (from simple line editors to complex programmable editors). I guess the best way to think about it is that it's not the equipment, it's the operator.

Having said this, I don't want to leave you with the idea that if you are happy with the editor you are currently using, then continue on with it; I feel that you should continually look at different editors and try them out. It never hurts to find something you like better than what you are currently using.

You should also be aware that different editors are better for different tasks. For example, I am writing this book using Microsoft Write. It has good editing capabilities and is WYSIWYG ("What You See Is What You Get"), which is appropriate for this task. For Assembler, I use an Emacs clone. For structured code, I use a programmable editor in which I've programmed in language constructs (i.e., "if () {} else {} /* endif */" for C). For e-mail, I use something completely different. I could probably standardize on one editor, but why? I'm happy and efficient with my mish-mash of different ones.

I should probably have said this before now, but your greatest resource for development tools is the Internet. Languages, simulators, editors, and version-support tools are all available for free (or at least demos are) along with the Microchip Tools. This will allow you to pick and choose what is best for you at minimal cost.

Languages

The PIC has been around long enough for there to be a number of different languages (and versions of each language) to choose from. In appendix D, I've listed a number of sources of languages that you can choose from. In this book, I will be writing the examples in Assembler, with pseudo-code to explain what the code is doing. Again, staying with the theme of this chapter, the choice of language is up to you; I feel it is important that you program in the language that you are most comfortable with.

Having said this, I still have some comments. There are a few features of a language implementation that I feel are important.

The first has to do with memory. All microcontrollers (and the PICs in particular) are not blessed with unlimited memory (or caching schemes that make it appear that way), either in terms of control store or variable RAM. The language and implementation that you use should be very frugal with its use of the memory resources.

Now, despite all this, one of the things that I have found in developing PIC applications is that a well-designed application does not require a lot of memory. In the applications that I have written for this book, you will see that none of them really come close to using up all the resources of the PIC, even though many of them are quite complex applications.

However, you are probably thinking that everything is written in Assembler, and almost by definition, Assembler only uses exactly what you require. Languages can use a lot of memory, especially if they aren't optimized. Before investing in a compiler, make sure you understand what type of code is produced and what is the typical amount of code you can use. Efficiency is measured in terms of execution time and program space. Poor compilers can be many times less efficient than good compilers.

Another aspect to how much Assembler code a compiler produces affects the efficiency of how you debug the code. When developing code for the IBM PC, I often find it useful to look at the produced Assembler code to see what is happening in a program that doesn't work. If the compiler doesn't produce efficient code, you might have a lot of problems understanding what the program is doing at a given point.

The next important aspect of the language is the data types used. Native PICs only run in 8-bit code; you should make sure the compiler that you use gives you a variety of data types (16-bit can be extremely useful). I've found that there are many times in which I would like to use more than 8 bits for counters and such. Many of the programs presented in here use 16-bit variables and, in the appendices, I've included a number of 16-bit Mathematical algorithms for your use.

Hardware support and initializations are another important aspect to look at when evaluating PIC compilers. The questions center around the question of what the compiler's initial code does before starting the application code (i.e., does it set certain features, such as timers and I/O ports, in specific states that may cause problems later?). If the compiler uses resources that you will want to use (i.e., the FSR register), you might have to change the way you were planning on developing an application because of the way the compiler uses and sets up the resources that are available within the PIC.

The next aspect is applicability across the whole PIC line. The language that you choose should produce code for all the devices in the PIC line-up. This is very important because you might be putting your application on a 16C84 for development and debugging, but your ultimate application might use a 16C54 (which is much cheaper). Using a compiler capable of producing code for all members of the PIC line means that new code doesn't have to be written when porting functions (or even whole applications) to different members of the PIC family.

Modifying the compiler-produced Assembly code is sub-optimal from the perspective of code support. Errors can creep in when changing code to support different PICs. These errors include having to change a number of locations, with some ending up being missed, or having to change code for different hardware implementations in different devices. Some programming techniques (such as only accessing specific hardware in subroutines) can minimize the opportunity for these errors in the future; however, if at all possible, the compiler you choose should be able to create code for different PIC devices.

Optimization of the produced code is an important aspect of the compiler. The compiler should be able to understand what is going on and use the most efficient code possible. The other aspect of optimization that you should be aware of is whether the compiler uses only the code that is required for the program. I personally like to have all my subroutines available in one file. Only the subroutines that a particular application calls should be included in the final object file. In the MPASM language, this is known as *conditional assembly*.

In much of the code given in this book, I have used conditional assembly to allow easy/fast debugging of code. Often, long delays are necessary when code is running in real-time. These delays can make the operation of the simulator/debugger very tedious and make it difficult to understand the flow of the program. By eliminating these blocks of code for debugging, the time required to execute the code is much less and it is much easier for you to follow the operation of the code.

A classic example of the need for conditional code is in code with long delays or complex interfacing requirements. During simulation, these portions of code can be easily skipped over using an externally defined label.

Typically, I use the "Debug" label like this:

```
ifndef Debug
  call UnreasonablyLongDlay ; Wait for external hardware
endif
```

When I am simulating/debugging this code, I assemble the code with "Debug" enabled from the command line (using the "/pDebug" option) and the "call" statement is not added to the code.

Even simple optimization routines can improve the speed of the compiled code by orders of magnitude and make a high-level language approximate the performance (speed and code size) of assembled code.

The last (and probably most important) piece of advice that I have about languages is: Don't pick one that hides the features of the microcontroller. When I say this, I mean that you should have direct access to all the PIC registers.

The reason for this, as paradoxically as it seems, is simplicity. If the language controls the interface to the hardware, you must learn how to use the language controls. This means that, on top of understanding the PIC, you also have to learn the language and its hardware interface. To make matters worse, the interface is designed by somebody else who doesn't have your application in mind; they've just written the most universally designed interface they could.

For example, the code to print "Hello World" and start a new Line would be:

```
? "Hello World": REM In BASIC
```

or

```
printf( "Hello World\n" );          /*  In "C"  */
```

Both "print" routines depend on specific hardware to function properly. If these functions are required for different hardware, it is important that you are able to access the required resources so that the necessary functions can be coded easily.

After reading everything discussed so far, you probably think Assembler is the best way to go. I think that high-level code development has significant advantages in terms of development effort and speed, and writing the code in a high-level language does prevent a lot of the typos endemic in Assembler coding. I'm trying to say that care should be taken to ensure that the language chosen should not limit what type of applications you can create.

One word on errors and warnings: I don't know if I'm just being anal retentive, but I don't like to use a program unless it compiles/assembles absolutely cleanly. This is because warnings have a way of becoming problems. Throughout chapter 8, I will point out errors and warnings, what they mean, and how to avoid them. The MPASM Assembler has a facility to allow you to disable certain messages, but I feel most comfortable with code that doesn't produce any messages of any type (other than "Everything's okay—go forth and execute!").

The PIC_II language

When I first started learning about PICs, I examined a number of languages available for the PIC and then decided to try to write my own. This exercise was to try to actually write a compiler (something I had never done before) and learn every aspect of the PIC that was possible.

The first version was a C-like language. While it did work well, I was not happy with the readability of the source. At the time, this had a lot to do with some prejudices against C and having to go through somebody else's code.

This lead me to the PIC_II language, which forced the user into one statement per line and avoided the need for C's braces and statement-ending semicolons. The language actually turned into a kind of pseudo-code.

For example, some C conditional code would look like:

```
if (( PORTB & 1 ) != 0 )  {          /*  Data available                  */
   a += b;
   PORTB = PORTB | 2;  Data++;        /*  Set the Acknowledge bit         */
} else                                /*  Data NOT available              */
   PORTB = PORTB & 0x0FD;
```

In the PIC_II Language, this would be:

```
if ( PORTB & 1 ) != 0               ;  Data available
   a = a + b
   PORTB = PORTB | 2                ;  Set the Acknowledge bit
   Data = Data + 1
else                                ;  Data NOT available
   PORTB = PORTB & 0x0FD
```

This seemed to be a much easier language to follow; the conditional nesting is carried out by the column the code is put into (when going "up" the nesting level, the code is moved back to the left to match with the appropriate code).

The language itself sets up a "pseudo-stack," which is used to temporarily store values during execution.

In chapter 9, I have included a project which uses the PIC_II language.

Looking back at the language, I can evaluate it using the following criteria:
- Readability
- Optimization/code speed
- Efficient use of code
- PIC resource availability
- Data structures
- Macro support
- Operability with other tools (i.e., MPLAB and MPSIM)
- PIC family support

When I evaluate the PIC language to this criteria, I get the following information:

Readability I personally think the source is very clear and readable. It might seem a bit tedious to code in, and sometimes having the correct column can be annoying.

Optimization/code speed The PIC_II language's optimization is as good as anything I saw. The compiler breaks down the source into *tokens*. The tokens consist of basically stack push/pop operations. Once the tokens have been created, the compiler looks for opportunities to optimize the tokens into simple pieces of code.

For example:

```
PORTB = PORTB | 2
```

from the previous code becomes the tokens:

```
push  PORTB
push          2
or
pop           PORTB
```

and is optimized into:

```
bsf           PORTB, 1
```

There are some compiler technologies that create code that will not make a lot of sense when you look at them (i.e., different approaches are used to carry out the requested functions, which use known internal conditions of the PIC). This might be an issue when debugging code if you're like me and like to look at the Assembler code to see exactly what is happening.

Efficient use of code One of the features I liked about this language is that there is no linking of object libraries. While some people might feel this is a disadvantage, I feel that this becomes a huge advantage when bringing in libraries.

Typically libraries consist of a number of compiled and linked routines. Even if you don't require a routine, it is still linked into your code (and taking up space). This can be a problem in something like a PIC, where the control store resources are limited. The PIC_II language eliminates all the routines that are not called from the mainline of the code. The mainline of the PIC_II language is simply anything that is not in a routine.

Also dealing with only source code has the advantage of being able to have conditionally executing constants and addresses, which tends to make the code smaller and more efficient.

PIC resource availability Rather than create a bunch of libraries that provide a standard interface to the various PIC resources, I let the user do it directly. Everything in the PIC, with the exception of the FSR register is available to the user both when programming the PIC and when programming in Assembler.

Data structures This section alone could fill an entire chapter of the book. When I discuss data structures, I am discussing being able to access data in the most efficient manner. The PIC_II language handles 8- and 16-bit numbers along with providing a Table Data Read Function; however, for the most part, I would give it failing grades because it doesn't handle arrays, pointers, individual bits, strings, or data types, and the Table Read Function only allows a maximum 251 instructions for all tables.

When I say "arrays" and "strings," I mean the ability to manipulate a series of bytes through simple commands. Bits should be manipulated directly, rather than through ANDing and ORing them. Finally, being able to define different data types as a block (which can be pointed to) makes handing packets of data in a specific format easier and more efficient. All these additional features would allow the language to handle much more varied tasks.

Macro support While I might appear to be down on macros in principle, I do use them when it makes sense. The PIC_II language treats both string replacement (i.e., defines) and line-replacement macros the same way along with the ability to pass parameters. An advantage of the macro processor is the taking on of "_xx" (where "xx" is the macro invocation) to each label, to allow the user to better follow what labels the macro is using.

Operability with other tools The ability to be able to use the PIC_II compiler with MPLAB would be a huge advantage. To do this, a .COD file must be produced.

PIC family support This is important because you might have routines (or even applications) that you would like to port to other PIC devices without having to change the code. The PIC_II language is very poor at this; actually it's really only designed for the 18-pin 16Cxx mid-range devices, which limits its usefulness to the 16C61, 16C71, and 16C84.

The previous criteria is more or less in order of what *I* consider to be important. The ratings I give are centered around what is important to me. When you evaluate a language, you might have different priorities, and these should be identified and categorized and then used when choosing a language.

You might find that no language absolutely meets all your requirements. However, by doing a careful, structured comparison, you will end up with the best match to your needs.

Version-support tools

Version support is a $10 phrase for keeping track of your software and making sure that the correct level is used and released for your applications. I find that, when I develop PIC applications (and any software in general), I tend to go through a number of different versions (starting with a hardware diagnostic going up to a functioning system) and experiment with different ways of doing things.

Sometimes, they're not always successful.

What I find that works for me is putting each application in a separate subdirectory on my PC. Then, I often start with a program named with version information and work my way up. In the comments, I note when this version is to be used and whether or not it is to be released.

As I say elsewhere in this chapter, sometimes I'm not always successful in keeping the programs up to date. To try and minimize this, if an experiment is not successful, I make sure it is clearly marked in the title bar and opening comments of the program.

While this isn't quite applicable to PIC tools at this time (there is no linker available for PIC object code), when I create PC applications, I write DOS batch files for creating the code. Each time I have something that works, the batch file is updated to link together only those object files that I know work properly. If a "Make" utility is available for the tools that I am using, I use this instead.

MPLAB helps facilitate version control through its use of projects. If you are not using MPLAB for developing your code, you might want to look around for version-controlling software that will do this for you.

Before you begin any application, the first thing that you should understand is how you are going to develop the code and how you are going to keep track of the different versions you create.

Simulators

There's going to be a time that you want to debug your program. A simulator is a piece of software that runs on your development system (PC) and allows you to watch your program execute. The major difference between simulators and emulators is what can be done in terms of debugging applications. The simulator is good for a gross-level debugging of programs; it is generally very difficult to get down to an individual cycle or hardware bit for debugging. The emulator can be used for complete debugging of an application, actually in the circuit, allowing for individual cycle debugging.

In this book, the Microchip free development tools will be referenced. This means that the programs will be debugged using MPSIM. To make using the simulator more efficient, I've included simulator set up (or "mpsim.ini" files) with each program. MP-SIM in MPLAB has watch windows that carry out the same function. As a rule of thumb, I like to be able to observe each register. The watch window and MPSIM.INI files I create allow me to do just that.

I am not comfortable using a simulator to check out a program to the individual cycle. To be able to do this, you must be able to control the phase of the clock going into the PIC. If this level of detail is required for the application development, then an emulator should be used.

There is one important point that I wanted to discuss in this chapter, and this is more of a philosophical statement. My personal goal is to develop software that doesn't require an emulator for debugging. To achieve this goal, the application should be reviewed and the design changed so that the timing is not critical to within several (at least four) cycles. In doing this, you will be almost totally eliminating any chance that the application will fail intermittently. When you are debugging, you will always prefer debugging solid failures to intermittent ones (trust me on this). Debugging will become more difficult as the period in between errors becomes longer.

MPLAB

Microchip has recently released a complete development suite for the PIC, called MPLAB. Contained within MPLAB is an editor, simulator, version-support tool, and interface to Microchip's PICStart Plus and PICMaster device programmers. MPLAB can also control the PICMaster emulator using the current source. Like MPASM and MPSIM, MPLAB is available free of charge from Microchip's BBS, ftp server, or Web site.

MPLAB can take MPASM source and process it into a hex file. The source is referenced within projects, allowing you to access only the code that is required for a given application (this is the version control). There are no differences in the MPASMWIN, which is used in MPLAB, to MPASM, which can be accessed from the DOS command line. It is important to ensure that MPASMWIN and MPASM are at the same level, if you are going to use both for developing/debugging applications.

Along with PIC Assembler (MPASM), other languages (such as C) are available for use with MPLAB.

The integrated simulator takes a bit of getting used to (especially if you have been working with MPSIM). The simulator (or emulator) uses the currently loaded (and assembled/compiled) file as source and is active as soon as the file (and its support files) are loaded. There is one major difference between MPSIM and the MPSIM (as the integrated simulator is known) and that's with regard to how registers used in the simulator are displayed. MPSIM uses an MPSIM.INI file to set up the simulator and list all the register addresses to be monitored (this is discussed later in this chapter). MPLAB allows the creation of watch windows, which gives you the ability to display the registers by variable name, rather than just address (as MPSIM requires you to do).

The MPLAB in-circuit emulator ("PICMaster") uses exactly the same commands as the simulator to eliminate any confusion and relearning of the two systems. As noted elsewhere, there are a plethora of programmers available, but only the Microchip PICStart Plus works directly MPLAB. These tools enhance MPLAB and make it an integrated application development system.

Going through the files

In MPASM, there are a number of files that you should be aware of and know about before you begin to develop programs or work the examples in this book. The files produced by MPASM provide data for the programmers as well as the simulator/emulator. I also want to make sure you understand what is required to develop the program.

MPASM (.ASM) source

The first file you have to worry about is your source code. In chapter 8, I will start the first few examples with the full code. To illustrate a point (feature or potential problem), I will just show the affected code. MPASM is relatively free of difficult constructs and control statements compared to other Assemblers. However, there is a layer of control statements that you should be aware of. These statements will be discussed later in this chapter.

Here is a sample of the body of the source file. This includes the requests for embedding other files into the code, the register definitions, and the actual code. After this, the last line has a simple "end" in it. This is used to tell the Assembler that the program is finished.

```
   title "Program Name - Brief Description"
;  Initial Comment Here           - Program name, author's name, dates, and
;                                           description

      :
   LIST P=16C84, R=DEC            ;  Specify options to be used for
                                     assembly
   INCLUDE "p16cxx.inc"           ;  Specify default labels and
                                     registers
Anything equ 12                   ;  Put in the register declarations
      .
      :
```

```
__CONFIG _CP_ON & _WDT_OFF          ;  Specify operating conditions of
                                       the PIC
org 0                               ;  Start the program at the reset
                                       vector
clrf   Anything                     ;  Initialize the variables
    .                               ;  Execute the program
    :
end
```

I keep to this format whenever I write a program. The information contained within here will help me find what I need to know when I come back to the program sometime later.

The "title" statement is used in the listing to display the name of the program. The string put here will be listed at the start of the listing. "Subtitles" can also be put in to identify different areas or blocks of code.

After the title, the program should begin with comments. The comments should point back to the name of the file, your name, your organization's name (and copyright statement, if required), the important dates of the program, and a brief description. All this information probably seems to be a waste of space, but you will find it useful (much) later when you have to debug the application or you want to take some sample code from it. Putting in the name of the file might seem somewhat (or very) anal retentive, but it's really for when I am editing multiple files at the same time and I want to know which one I am in. I also find putting in a simple revision history to be extremely helpful as well. I'm not great at updating the code to mark what I have changed, but listing out at the program start what has been done usually helps a lot. Well-done and consistent initial comments will help you find old files and include members that will help you later. You should experiment with the format that works best for you; in the examples used in this book, you will see my format and you can branch off from there.

An important item that should be included in the initial comments of every PIC program is a list of PIC pin information and uses. This will give you easy to find documentation of how the PIC is supposed to be wired.

Following the initial comments, the first Assembler command is the "LIST" statement. The "LIST" statement allows you to set the initial operating conditions for the Assembler. In the "LIST" statement, you will be specifying the processor to be used, the default radix to be used by the program, and the output file format.

Figure 6-1 lists what I consider to be the most important assembly command parameters.

These parameters can either be specified from the DOS command line as:

```
MPASM [/option[, /option...]] Source.ext
```

or in the LIST Statement of the source.

Personally, I prefer putting all the options inside the source code in the "LIST" statement. This allows you to consistently assemble the code, without worrying about making sure the correct parameters are specified in the command-line assembly statement. The only command-line parameter that I change regularly is the "Define Symbol" to specify "Debug" used when debugging an application.

Now, I'll make just a few comments on the assembly parameters. You will see throughout the book that I specify a base 10 (decimal) default number base system

Command	Format	Default
Display Help	/?|h	
Set Hex File Fmt	/aINHX8M|INHX8S|INHX32	/aINHX8M
Define Symbol	/dDebug /dNum=3 /dStr="12"	
Enable Macro Expansion	/m+ /m-	/m+ (On)
Set Processor Type	/p16C84	NONE
Set Radix	/rHEX /rDEC /rOCT	/rHEX

6-1 Command-line options.

(if the parameter is not used, the default is hex). I do this because I think in base 10 and hex values can be specified easily (using the "$0x0xx$" format). This is the major deviation I make from the defaults specified in the program. I really recommend staying with the default parameters, as such things as hardware and software (i.e., programmers, emulators, compilers, etc.) are generally designed around keeping to these formats.

Following the LIST command, I put in all the standard register and macro declaration files. In the example body, you can see that I use a standard file provided by Microchip. Wherever possible, I try to use the standard tools available. There are two reasons for this. The first is that I want to be as standard as possible (the standard register declarations use the Microchip naming conventions). Because, if I do have a problem and have to go and request help, I want to have something that other people can recognize easily. The second reason for using standard declares is that I am lazy. Actually, the latter reason is more important than the former. I find that letting other people do the work wherever possible makes me more productive.

After the standard definition file loads, I declare all the RAM registers that are required for the program. I try to keep the list simple, with the only wrinkle being, if I'm going to use the FSR register to access RAM, I put the memory for this after the specific register declarations. The reason for doing this is to simply ensure that, if the FSR controlling software has errors and FSR leaves the expected block of file registers, I can see this easily during simulation.

The __CONFIG statement can be used to control the configuration fuses of the PIC. These "fuses" are actually a register located at address 0x02007 in the address space (higher than most PICs can access) for the low-end and mid-range devices. For the high-end devices, the configuration fuses are at address 0x0FFF0. These "fuses" tell the PIC how to behave on power-up in terms of what type of oscillator is to be used, whether or not the watchdog timer is going to execute, etc.

I find one of the best ways of ensuring the PIC can be programmed properly is to include the configuration information in the source (similar to one "LIST" statement), rather than entering it in each time before programming the devices. The configuration fuses available are specific (in terms of feature, location, and polarity) to each PIC. However, in general, the fuses to be most concerned with (and are available in all PICs) are those listed in Table 6-1.

Table 6-1. The configuration fuses

Fuse	Description
_OSC	Type of clocking used for the PIC
_WDT	Watchdog timer enable
_CP	Code protect enable

There are also a number of other fuses specific to the different PICs. These include those listed in Table 6-2.

Table 6-2. Fuses specific to the different PICs

Fuse	Description
_PWRTE	Power-up timer enable (only available in mid-range)
_CP_Fraction	Partial code protection schemes
PIC Mem Model	Specific to 17C4x devices

One thing to note with the __CONFIG statement (and __IDLOCS) is that the command cannot start in the first column of the line. The first column is reserved for labels, and putting these commands on the first line will confuse MPASM.

The __CONFIG data cannot be read by the PIC itself except for the high-end devices.

A word of caution with regard to the use of the "Code Protect" ("__CP") bit(s) of the configuration fuses: This bit should only be made active (i.e., the code cannot be read from the PIC) if you consider the code complete, debugged, qualified, certified, and generally perfect in every respect. When this bit is set, you will be unable to read back the program memory. This might not seem so bad; however, in the latest PIC devices, the Code Protect bit cannot be reset after device erasure, and you will not be able to either erase the PIC or to load in another program. The code protection should only be enabled when you are absolutely sure you want it to be used.

Along with the configuration fuses are four registers in the low-end and mid-range PICs (the high-range doesn't have this feature) used to store information external to the PIC. This information might be device (or product) serial numbers or code-level information. Each register has 4 bits available for this purpose. To specify these values, the __IDLOCS command is used with a 16-bit constant following (like __CONFIG, __IDLOCS cannot start in the first column). The 16-bit constant is broken up into four 4-bit words, each of which is put into a separate IDLOCS location.

If __IDLOCS is not explicitly stated in the source, and at programming, the Microchip programming software will put in a default checksum value. This checksum is calculated from the object code to be loaded into the PIC. Different programmers and software might use a different checksum algorithm from the Microchip standard (resulting in different values being put into the code), which might cause miscompares when programmed parts are verified on different programmers. These miscompares can be ignored without worrying about how the program will execute.

I find it best to not use the __IDLOCS command and let the programmer calculate a value to be put in there. This value can be read back through the programmer and compared against an expected value (giving a small amount of code-level control).

The "org 0" statement tells the Assembler to begin the code at address 0. In all the PICs except for the low-end devices, the address the PIC starts executing at is 0. For the low-end PICs, the first address the PIC begins to execute at is the last address in the program memory. The first statement after reset in a low-end PIC should always be a "goto" to the start of the program.

Often, you'll want to develop code for a low-end device on a 16F84 (because it's EEPROM allows you to skip the erase step). When doing this, you should put in two "goto Mainline" statements in the code: one for the 16F84 and one for the 16C5x device. This could be done as a conditional assembly statement (described later in this chapter); however, by putting both statements in, you have a check that will cause an error if your code goes beyond the 16C5x end of program memory (there will be a "Program Memory Overwritten" error).

Additional "org" statements might be required (for the interrupt handler, which starts at address 4, or table routines, which are often easier to do if they don't cross a 256 address boundary—this is explained in chapters 2 and 8.

Before starting the program, I always make sure that I initialize all the variables that I use. This is important because, in most simulators (including the Microchip tools), the initial state for all registers is zero (0x000).

This is *not* the case in the PIC. RAM registers can be any value. To make sure I don't get into any problems with what the values actually are, I initialize them. If I don't require them to be any value, I will sometimes set them to zero (using a "clrf" instruction) to make sure the registers are exactly what's shown. When you run MPSIM, you will be notified if a register is read before it is written. When you receive this message, go back and find out what is happening; I've been in trouble several times by not heeding this message.

Labels and comment lines should start in the first column of the line. MPASM will accept labels anywhere in line, but MPASMWIN (the Windows version of MPASM used in MPLAB) will not. Starting your labels and comment lines in the first column of your code makes for a good visual cue for your software.

This:

```
Label
        ;  Execute Only this code
```

looks a lot better as:

```
Label
;  Execute Only this code
```

Comments are preceded by a semicolon. Lines can start with a semicolon, or the comment can start wherever on a given line. Often when debugging programs, I will turn off a section of code by putting a semicolon at the start of each line. This is known as *commenting out* the lines. By putting a semicolon at the start of a line, everything following on that line will be ignored when you assemble the program. This is a simple way of editing your program without loosing the code.

Lastly, comments are meant to enhance the understanding of the program. They should not be used to try and explain everything that is going on in the program. Care should be taken to make sure that your comments don't restate the obvious. An example of a needless comment is:

```
incf   i, f          ; Increment "i" and store the result back in "i"
```

Saying that I am incrementing a value isn't required when the instruction does it for me.

One thing that you will see in all the software in this book is that I always make sure the program is explicitly ended. This means that, when I have completed everything that I want to do, I either put the PIC to sleep (simply by executing the "sleep" instruction) or put the PIC into an endless loop (perhaps waiting for input). Usually the endless loop is the preferable choice because the I/O state of the device cannot change (as it can with the "sleep" command), or it is in a base state waiting for external input to execute a response.

As noted above, a tool to simplify and make your assembly language coding more efficient and readable is *conditional assembly*. This is where code is only added to the program when certain parameters are met.

For example, if you wanted to put in different paged "gotos" for different devices, you could create three separate include files, one for reach type of PIC. You also could use a single file with conditional assembly.

With conditional assembly, the "gotos" could be handled like in Fig. 6-2. Only the correct "mgoto" macro will be entered into the code.

As you can see, the "ifs" (for conditional assembly) can be nested to quite a reasonable level (three levels in the example in Fig. 6-2). The "if" checks a condition directly, and the "ifdef" checks to see if a label has been defined or exists by the "LIST" statement (for PIC type) or a "define" statement.

Here's one hint with conditional assembly that I find makes my debugging easier. When you are running a simulator on your code, you will find that delay loops that are quite short when the PIC is running (i.e., a 10-msec delay) might take literally hours when you are debugging code in the simulator. For this reason, I often surround delay loops with an "if" looking for a "Debug" defined label:

```
ifndef Debug                    ; Execute only if "Debug" is NOT defined
  call          Dlay            ; 1/2 second delay
endif
```

This method will make your debugging quite a bit easier and will allow you to switch between Debug and Running code very easily without having to resort to commenting lines of code out (and subsequently forgetting what you've commented out and why).

When MPASM starts to run, the processor type declared in the "LIST" statement or command line is used to create a define label. The define label is the PIC type preceded by two underscores (therefore for a PIC 16C64 has a define label of __16C64). The conditional code discovers what range of PIC the current device fits into and creates an appropriate long "goto." Note that I have indented the nested "gotos" in Fig. 6-2 to keep their nesting level straight and visually obvious.

6-2 Handling the "gotos" with conditional assembly.

```
     ifdef   __16C54          ; Define a variable for low-range PICs
#define  lowrange
     .                        ; And so on for each device
     :

     ifdef   __17C42          ; Define variable for high-range PICs
#define  hirange
     .                        ; And so on for each device
     :

     ifdef   lowrange         ; Put in the goto for low-end PICs

mgoto  MACRO  Label
       movlw  0x09F
       andwf  STATUS
    if ( Label & 0x0600 ) != 0
       movlw  ( Label > 5 ) & 0x060
       iorwf  STATUS
    endif
       goto  Label & 0x01FF
       endm

    else                      ; Else, look for high-range PICs
       ifdef hirange

mgoto  MACRO  Label
       movlw  high Label
       movwf  PCLATH
       movlw  Label & 0x0FF
       movwf  PCL
       endm

    else                      ; Else, left with mid-range devices

mgoto  MACRO  Label
       if ( Label & 0x01800 ) != 0
       movlw  high Label
       movwf  PCLATH
       else
       clrf  PCLATH
       endif
       goto  Label & 0x07FF
       endm

    endif
  endif
```

Once your source is complete, the last line is an "end" statement. This statement tells MPASM that the program is complete. For this reason, I really don't bother with ending comments. Because I use the "title" statement to identify which file I have, when I look at a listing, I know what the "end" is for. I do not put "end" statements in included files for this reason.

Standard declaration (.INC) files

With the Microchip MPASM software comes the standard register declarations for the different devices. In the software presented here, I use the standard P16C*xx*.INC file (where "xx" is specific to the PIC used), which declares the various registers used by the device as is described in the Microchip datasheets. Along with the Microchip registers, you might want to declare your own specific registers or load in your own macro files. This can be done using the "INCLUDE" statement. Code in the "INCLUDE" statement is written exactly as you would any other code.

Listing (.LST) file

The listing file will give you a complete picture of how your code was assembled or compiled and turned into PIC program data. The file consists of each line of code used (and not blocked by "NOLIST") along with the value and address (if appropriate), the variables used, and the control store used.

The listing file shows how a program was assembled, along with addresses and actual code. Also included is a list of labels used (as I've pointed out before, MPASM does not assign variables, it allows you to use labels in the register space); a list of errors, warnings, and messages; and a map showing how the program will be loaded into the PIC's program storage space. A quick glance at the program storage map can identify problems before you even begin simulations or programming.

In MPSIM, I use this mostly as a debugging tool to follow what is happening when I am debugging an application using the simulator. This isn't necessary in MPLAB, because program and variable addresses are communicated directly between the Assembler (or compiler) and the integrated simulator.

I don't tend to keep the listing file (the source files being much more important), but it is useful for following how the code is executing for the DOS command-line simulator and understanding where the code is put into the PIC.

Error (.ERR) file

The error file stores all the errors, warnings, and messages produced by MPASM. In MPLAB, this file is displayed after program assembly if there are any errors in the code and allows you to go directly to the failing line. When working from the command line and using MPASM/MPSIM, I really just look for a .ERR file length of zero (which indicates there weren't any problems with the assembly).

Errors, warnings, and messages can be suppressed by the "errorlevel" command:

```
errorlevel 0, -305        ; No Messages for missing ", f"
```

However, I use the "errorlevel" command as little as possible to make sure I truly understand how the code is working and any issues encountered during assembly.

Symbol (.COD) file

The symbol file is used by the simulators and emulators to work with source code directly with its symbols (i.e., labels and variable names). This is a file that is largely transparent to you, but make sure that, if you are debugging a file using MPSIM, you have it available along with the listing and source.

Object (.HEX) file

The last and probably most important file output from MPASM is the object or .HEX file. This file contains all the information for programming the PIC. The .HEX file will only be produced if the source code assembles without any errors. However, as noted previously, in the .ERR file description, errors can be blocked and the .HEX file will be produced regardless of any problems.

The format of the object file is one of three formats. The default format is "INHX8S," which is a standard Intel "Intellec" hex format. This format is used by most available programmers and should *only* be changed if the programmer you are using specifies a different format.

The programmer and emulator provided with this book use the standard ("INHX8S") format.

MPSIM.INI

The MPSIM.INI file will be one of the most useful files that you can create. It is used to control how the MPSIM simulator starts up when you are debugging your program. Such things as the target PIC, the default radix for values, the registers to monitor, and the clock period you are using can be set into the MPSIM.INI. Along with source, all the programs that I give in this book will have an MPSIM.INI to work with it.

The simulator (MPSIM) does not have a great method of displaying data; you will probably have to work with it for awhile to understand what it is trying to show you. Some patience and perseverance is required (you might want to download MPAB from the Microchip Web site rather than use the copy of MPSIM provided in the book for this reason). Please don't think that the method I use for displaying registers is the only one available to you; you can develop your own conventions that make the most sense to you.

Figure 6-3 shows a typical MPSIM.INI file.

An important point to note in this file is the use of the comments and the absence of "white space" to indicate breaks in the different areas of the file. White space, or blank lines, cannot be put into the MPSIM.INI file because, at each blank line, the previous instruction will be repeated. The commands and their various parameters are explained in Microchip's "MPSIM User's Guide."

The first part of the MPSIM.INI file is used to set the simulator up for the processor and the basic defaults. Register data makes sense to me in hex (because we are talking about 8-bit registers). This might come in conflict with what I have said elsewhere, but I do prefer looking at data in registers in a manner in which I can convert to binary (to understand the bit pattern). This is not to say that I don't use binary for some registers (in which each bit is a separate entity) or decimal for constants. I use the data display format that makes the most sense to me.

I have tried to come up with a standard format for how I work with the simulator. This means that I can always be pretty sure of where to look for various registers. The best analogy I can come up with to illustrate how I do the registers is in terms of an airplane cockpit. There are six basic instruments that every aircraft has had since the Second World War; they are all clustered in the middle of the cockpit for the pi-

6-3 A typical MPSIM.INI file.

```
;   MPSIM File for PROG2 - Turning on an LED
;
;   Myke Predko - 96.05.20
;
P 84                      ;  Use a 16C84
SR X                      ;  Hex numbers in the simulator
ZR                        ;  Zero the registers
RE                        ;  Reset elapsed time and step count
DW D                      ;  Disable the wdt
V W,X,2                   ;  Display: the "w" register
AD F3,B,8                 ;     STATUS register
AD F4,X,2                 ;     FSR register
AD OPT,X,2                ;     OPTION register
AD FB,B,8                 ;     INTCON register
AD F2,X,3                 ;     PCL register
AD FA,X,3                 ;     PCLATH register
AD F1,X,2                 ;     TMR0 register
AD IOA,X,2                ;     Port "A" TRIS register
AD F5,X,2                 ;     Port "A" register
AD IOB,X,2                ;     Port "B" TRIS register
AD F6,X,2                 ;     Port "B" register
AD FC,X,2                 ;     "Test" register
rs
sc 4                      ;  Set the clock to 1MHz
lo prog2
di 0,0                    ;  Display the first instruction
```

lot to find easily, without much scanning. I try to do the same thing with the registers; I put the primary execution registers first. They include:

- The "w" register
- The STATUS register (This is put in binary format.)
- The Program Counter
- The FSR register
- Port "A" I/O control (TRIS) and data registers
- Port "B" I/O control and data registers

Looking at the MPSIM.INI file listing in Fig. 6-3, you can see these registers are specified first (which will put them at the top of the MPSIM screen), which will make them easy to find.

After these basic registers, I then include any specific hardware control registers that are used by the program (i.e., INTCON or TMR0).

Following the hardware registers, I put in the specified variable registers. To debug a program using MPSIM, I usually print out a copy of the listing file and use it to follow the action. For this reason, I make sure that I display the variable registers in the same order as what's in the program.

As will be seen, the labelling on the registers leaves much to be desired. The conventions used here for defining where the registers are placed does help make understanding the data displayed easier.

Once all the registers are defined, then the basic setup of the program, including specifying the clock period of the program, is put in the MPSIM.INI file. Finally, a command to load in the program is put into the MPSIM.INI file.

You can put any valid MPSIM command into the MPSIM.INI file. When I'm involved in heavy debug, I'll often update the file to include breakpoint definitions. While MPSIM is not great at showing you what's going on, a well thought out MPSIM.INI will make your application debug much easier.

As noted previously, if MPLAB is used for its integrated development system and simulator, you will probably want to set up a custom watch window for the registers that you use in your application. This is analogous to the MPSIM.INI file's register specifications. One thing to note is, in MPLAB, all possible versions of the data (decimal, hex, binary, and character) are shown, you don't have to worry about specifying the actual value.

A watch window would probably look something like Fig. 6-4.

6-4 A watch window.

Symbol	Hex	Dec	Binary	Char
w	00	0	00000000	.
STATUS	18	24	00011000	.
PCHI	00	0	00000000	.
PClo	00	0	00000000	.
INSHI	00	0	00000000	.
INSlo	00	0	00000000	.
DataHI	00	0	00000000	.
Datalo	00	0	00000000	.
i	00	0	00000000	.
TXOut	00	0	00000000	.
RXIn	00	0	00000000	.
PCLATH	00	0	00000000	.

Stimulus (.STI) files

For programs that require input from the hardware around it, you'll need a stimulus file in order to debug it in the simulator before burning the chip. As I'll note later in this chapter; the most significant PIC debugging happens at the PC, not the workbench. PIC input can be injected asynchronously into the simulator as the program is executing, but a stimulus file offers the advantage of being able to offer the PIC a consistent data input for debugging.

Stimulus files allow you to define what the program is going to encounter in terms of input. Stimulus files generally allow you to define waveforms and conditions that the PIC will encounter in the application.

The format of the stimulus file is actually quite simple. Input value changes are made at specific step counts. This means that the pins to be input have to be identified along with the step count. The same stimulus file format works for both MPSIM and MP-SIM. However, there is one significant difference that is described later.

All input bits of the PIC can be referenced. This is done by a pin identifier. The reset bit is referenced by the label "MCLR." Port input bits start with an "R" followed by the port identifier letter and finally the bit. So Port "B" bit 4 would have the identifier "RB4."

Figure 6-5 shows a sample stimulus file (which, by convention, always ends in ".sti"):

6-5 A sample stimulus file.

```
!
!  Sample Stimulus File
!
Step      MCLR      RB4       !  Define the bits to be controlled
  1         1         1        !  Initialize the bit values
!  Wait for the program and hardware to be initialized
  500       0         1        !  Reset the PIC
1000        1         1
1500        1         0        !  Change the state of the port bit
2000        1         1        !  Restore it for rest of program
```

The lines beginning with a "!" are comments. The "!" character is used to indicate that comments follow to the end of the line. The first actual line is the "Step...," which is used to declare the bits that are controlled by this stimulus file and the order in which they will be presented.

The following, noncommented lines are the input data to the simulated PIC. As indicated previously, the signal value is asserted at the cycle count value. This means that some latency issues cannot be explored by the simulator.

The MPSIM stimulus files are easy to develop, but there is one thing to note: They are not time-based. This means you have to guess at the instruction count unless you set the simulator for stepping by time (using the "ip time" command). In some of the applications, complex timings (i.e., the I/R receiver and keyboard input) took quite a while to set up correctly.

However, they saved me many hours of debugging the hardware.

To determine the "Cycle Step Count" for a given time, the following formula will give the number of cycles for a specific time delay:

$$step = \frac{(time\_dlay * frequency)}{4}$$

This formula finds the number of oscillator clock cycles to get the time (time * freq) and is then divided by four to get the actual number of instruction cycles.

For example, calculating the number of cycle steps for 20 msecs, in a PIC application with a 10-MHz clock yields a step count of:

$$step = (20 \text{ msec} * 10 \text{ MHz}) / 4$$

making the units consistent:

$$step = \frac{(20(10^{**}-3) \text{ sec} * 10(10^{**}6) \text{ 1/sec})}{4}$$

$$= \frac{20(10**3)}{4}$$
$$= 5(10**3)$$

The cycle step count at 20 msec in this example is 5000. In the stimulus files, the step counts are absolute, so the cycle count should be added to the last step value.

You'll feel as if you are going blind as you debug programs with stimulus, but the effort is well worth it when it comes to the final product and the program you burn into your PIC works first time.

MPLAB files

MPLAB produces a number of files that are not common to the DOS command-line version of the assembler and simulator. These files are largely used by the project (version-control tool) of MPLAB. These files are a by-product of using the mouse to set up the screen and display optional registers as the user likes to see the data presented.

The .PJT and .CFG files are the control files for the project. The files contain a list of all the files associated with the project, what windows were open when the project was saved, and the window's location. These files, while readable really won't tell you a lot.

Going along with the project files is the watch window file. As discussed earlier (in the MPSIM.INI section), these files (ending in ".wat") are created during debug of the application to watch different registers. I usually set up all the registers that are used in the program so that I can watch how the code is executing. The watch window is the closest analogy to the MPSIM.INI file; it is used to define the registers to be watched while the program is being simulated. In chapter 8, I will show how the watch window is set up.

The final two files that only appear in MPLAB are the .$$$ and .BKX files. These files are used to save (back up) the most recent copies of the previous source and hex files. Like all back up files, these should only be used for emergencies and mistakes (like accidentally saving an error).

All the files described in the previous sections are in addition to the files listed previously and are really not meant to be accessed by the application developer.

Programming styles

Whole books have been written on what is the correct way to write computer code. I'm not going to try and replicate all that work here, but I will have a few comments to make about writing readable (and, therefore, easily debuggable) code for the PIC.

For this book, all the examples given will be in Assembler. A lot of the concepts presented here are also applicable to high-level languages.

Having said all this, one of the first things I'm going to do is recommend that you look for a high-level language for continuing your work on. In appendix D, you will see a number of sources of compilers for PIC C, Forth, Basic, etc.

Despite the comments and requirements I've given earlier, I do believe that the best software is written using high-level languages. I don't like wading through Assembler; I don't think you should do that to yourself if there are better alternatives either.

However, I should point out that I develop pseudo-code to explain what is happening, and the pseudo-code can be embedded into the Assembler code so that you can understand what the Assembler is trying to accomplish. My Assembler is developed with knowledge that a compiler couldn't have, so if the same code was converted to a language and compiled, chances are it wouldn't be as efficient as the Assembler source (although you might disagree with this statement after you've seen my code).

A good rule of thumb that I follow is that compiled code should take up no more than 25% more space or take more than 25% more cycles than well-written Assembler. After going through this book, you will be able to create efficient Assembler, and you will be able to judge the quality of the code produced by a compiler by yourself.

Probably the best piece of advice I can give you regarding developing code is: Copy, steal, and plan for the future. This means that the most successful programmers re-use what they've already developed, understood what other people have done and imported it into their projects, and, if they couldn't do that, they developed code with the idea in mind they would probably need to re-use it later, so they develop it to be usable in a variety of different situations.

Copying code means that you should make sure you control and save the code you develop in a manner in which you can find it again easily. I touch upon this a bit earlier in this chapter, in terms of code control, but you should make sure that you save everything that you write and make sure you can get back to it in some way in the future. Not only will you find bits and pieces that you will be able to copy directly into your new application, but you will find ways of doing things (that is in your own way of idiomatically writing), eliminating the need to search through a databook and relearn how to do something.

When I say "steal," I don't mean it in the literal sense. I mean for you to keep a watchful eye on what's out there and understand how other applications carry out various tasks. In chapter 9, I will present an infra-red-controlled tank. The genesis of the project was an article I read in *Electronics Now* for an I/R-controlled robot. The code that is presented is a complete rewrite that I have done for the code required to understand how the I/R receiver works as well as driving the electric motors. While I didn't use any code from the article, the understanding that I received from it was invaluable in making my own application.

If for some unfortunate reason, you end up having to develop code on your own, remember that you should be doing it with an eye toward using it again in the future. Doing this is actually very simple. The most important aspect of the development effort is to make sure that what you do isn't limited or that the limits are documented in the source. This source can be put into a *library*, which is a link with unique application code. Actually, a library is usually just a bunch of routines that appeared usable in other applications by the developer.

One word about libraries and code control: If anybody is working with you on developing code, don't count on your libraries being at the correct level; always go back and make sure that what you have is correct. Included in this topic is the idea of linking. Currently, MPASM (and most other PIC languages) do not support linking modules together. This means that all the code must be present at assembly/compile time. Personally I think this is a good thing, because it ensures that the source code you are using is at least at the latest level that *you* have.

One of the basic tenants of code development that I live by is the use of top-down programming strategies with device drivers, or subroutines carrying out the hardware interfaces that are required. This aids in debugging considerably, allowing me to jump through code quickly to find the problems, as well as allowing me to find and put a breakpoint at the function required for debugging in the simulator easily.

As noted earlier in the book, the destination of a number of instructions is specified at the end of the Assembler statement. For low-end and mid-range PICs, the destination is either the "w" register or back into the source register. If this is left out in the instruction, the default is storing the result back into the destination.

Thus, I find that an instruction like:

```
incf        i, f                ;  Increment "i" and store back in "i"
```

is much more readable as simply:

```
incf        i
```

Whitespace is the term for leaving blank lines in your program to mark different sections or functions of code. In the examples, you'll note that I leave spaces to de-mark the start and end of different functions. For example, if you were incrementing an 8-bit variable and adding it to a 16-bit variable, you could do it as:

```
incf        i, f                ;  Increment the first value
movf        i, w                ;  Add "i" to 16-bit "j"
addwf       j, f
btfsc       STATUS, C           ;  Do we have to increment the high byte?
incf        j + 1, f            ;  Yes
```

This can be made more readable by putting in a blank line between increment-ing the first value and the addition to the second along with removing the destination when the result is put back into the source:

```
incf        i                   ;  Increment the first value

movf        i, w                ;  Add "i" to 16-bit "j"
addwf       j
btfsc       STATUS, C           ;  Do we have to increment the high byte?
incf        j + 1               ;  Yes
```

Another tool to make the code easier to read is by adjusting the indentation to make note of conditional code. This is something that I have taken from structured languages. If you indent code that is conditional upon a previous line, you will also be able to pick out easily what the conditional code is. This is actually made simpler by the PIC because of its skip on condition instructions (the code that is not skipped is what is indented).

For the previous example, the conditional code (the incrementing of the high byte of the 16-bit variable) can be made more noticeable by indenting it relative to the other code:

```
incf        i                   ;  Increment the first value

movf        i, w                ;  Add "i" to 16-bit "j"
addwf       j
```

```
btfsc       STATUS, C       ;  Do we have to increment the high byte?
   incf     j + 1           ;  Yes
```

By using just meaningful comments, this block of code becomes:

```
incf        i

movf        i, w            ;  Add "i" to 16-bit "j"
addwf       j
btfsc       STATUS, C
   incf     j + 1
```

So, by paying attention to how the code is written, you can significantly improve the readability of your software and also make it easier for you to understand and follow. You should also note that a lot fewer keystrokes are required to be input, resulting in faster code development.

For some reason, I have a lot of emotion when it comes to macros. Actually, it's not for some reason, it's for *one* reason. Back, years ago when I first started working, I was given the task of finding a bug in a proprietary real-time operating system. The bug was pretty simple, easily repeatable, and I had a pretty good idea of what I would have to do to find it and fix it. I then started going through the source code and found that the source consisted totally of macro invocations. There were macros to do literally everything in the operating system. Finding and killing this bug, which I had confidently proclaimed would only take me a week, ended up taking me almost two months.

In case you don't know, macros are simple blocks of code that are created and invoked in a manner that is similar to a subroutine. The difference between a subroutine and a macro is that the code for the subroutine is located elsewhere and the macro is replaced by the code it represents.

I feel that macros are best suited for small pieces of code that are often repeated and using the macro makes the development of code easier. The reason for all the macro code in the operating system was a way in which the authors of the operating system could ensure that any changes to the code would be propagated throughout. Personally, I think there are easier ways of doing this (i.e., commonly loaded files), but then I'm not a PhD. like the authors of the operating system.

The PIC microcontrollers have a few operations that are not intuitive right off the bat. One of the first areas that I've found this is in conditional branching. For example, to jump on not zero, the Intel i86 (which powers the PC) uses the instruction:

```
jnz label
```

In the PIC, to do the same thing, the following two instructions are required:

```
btfss status, z
 goto label
```

This is obviously a lot more complex than the i86 example; however, with a macro, I can replace the two instructions with a single macro that does the same thing. The following is the final macro, which will look exactly like the i86 instruction when it is inline with the source:

```
jnz         Macro       Label
            btfss       STATUS, Z
             goto       Label
            endm
```

Macros are also useful in doing repetitive functions that are unique to the application. For example, say you were writing to a device that has a strobe. Rather than requiring the following code:

```
movf        DataOut, w
movwf       portb           ;  Put the data out on the data bus
bcf         porta, 0        ;  Pulse the strobe bit
bsf         porta, 0
```

to be replicated, this code can be put into and used as a macro.

Now, the question that you might be asking is, "Why would I use a macro over a subroutine?" The answer to that is pretty simple. It has to do with timing. For timing critical sections of code, a macro is preferable because calling a subroutine requires a minimum of for additional cycles (two for each the call and the return statements).

Just as something you might be interested in, here is an interesting macro that will swap the contents of "w" with the contents of another register without using an intermediate register (as is the normal algorithm). If you don't believe it works, put it into a program and run it through a simulator:

```
swapRegw        MACRO       reg
                xorwf       reg, w
                xorwf       reg
                xorwf       reg, w
                endm
```

I write lousy comments. For some reason, when I write code, I can't explain what I'm trying to do in plain English. For high-level code, this isn't a problem because the purpose of clearly written code is obvious. However, for Assembler, this becomes a big problem.

Some people like to write a detailed description of what the code does at the top. I try to do this, but I'm not great at keeping it up-to-date and understandable. What works for some people when they are writing assembler is to write out what the code is to do in pseudo-code and then write Assembler around it. Just by looking at the commented pseudo-code gives an idea of how a program is supposed to work, and changes are really pretty obvious (although you should still write a comment or two to explain what my thinking was).

Something that goes hand in hand with comments is constants. Constants should be descriptive or not used at all. If you want to AND a value to clear specific bits, I find using the bit string b'11101110' a lot more descriptive than a constant labelled, "clear_bit_0and4."

As a note, in chapter 8, you won't see a lot of pseudo-code. This is because a lot of the software is used to explain and demonstrate features and methods of the PIC. As you develop your own applications for the PIC, you will get a better idea of how things should be done.

Before starting a PIC Assembler program, one thing that you should decide is how to handle PCLATH (if changing it is required in the PIC you are using). You might want to keep the code very small and stay in one code page. If you can't do this, there are two basic methods that can be used. The first is keeping functional blocks of code in one page and not changing PCLATH until execution jumps to a different block. The second is to always update PCLATH for each and every time the Program Counter is changed (i.e., "goto" or "call") in the program.

The first method of handling PCLATH can be considered to be a very structured method of doing this. (See Fig. 6-6.)

6-6 The first method of handling PCLATH.

```
org 0                    ; First page of code

.
:
movlw    HIGH IORoutine  ; Set up PCLATH with the correct high
movwf    PCLATH          ; Order address
call     IORoutine
clrf     PCLATH          ; Restore PCLATH.
.
:

org 0x0800               ; High page of code

IORoutine
.
:
return
```

In Fig. 6-6, you can see how functional blocks are kept separate from one another and that there is no danger that code from different functions can be accidentally invoked. Code space and execution time is reduced because PCLATH is only changed upon entry into each block. In the interrupt handler, PCLATH must be saved before it is changed into the device handler page. Most importantly, this method makes it easier for you to write the code; you don't have to worry about PCLATH until you are leaving the functional block of code.

The second method, updating PCLATH each and every time the Program Counter is changed, can be simplified by using macros for "goto" and "call" (examples are earlier in this chapter in the "MPASM source" section). This method is attractive because it makes the PIC look like a completely "flat" memory device, rather than a bunch of blocks. (In the PC world, Intel was flooded with requests for a flat memory model for the $80x86$ devices and finally provided it in the '386.) The disadvantages of this method are that macros must be used instead of native instructions, and the code produced takes up more space and cycles than the single address used by the native instructions.

Note that, in both these methods, the contents of the "w" register is lost when moving between the pages. This means that the parameters that are normally passed in "w" have to be saved in temporary variables when calling routines that pass between pages.

I've never seen version 1.00 of a program that worked properly. This means that, after you write your code, you should review it and look for areas that it can be improved upon. Because the PIC (and other microcontrollers) have limited program storage capacity and a clock speed that is fixed (as opposed to a work station or PC that doubles speed and capacity every couple of years or so), it's vitally important that you continually look for areas that can be cut down or improved upon.

Going further in this, I generally develop code in a series of steps, each one differentiated in how it accesses external hardware or processes information. This method of program development can be seen throughout the iterations of source code in the projects supplied with the book. Carrying out development in this manner simplifies application debugging because, before going on to the next step of the development, the code is properly debugged. If problems surface in later versions of the code, it is easy to go back and rework the previous versions to understand the problem and fix it.

In writing any type of code, the one lesson I have learned over the years is be consistent. It really doesn't matter if your code conforms to somebody else's "ideal" as long as you do things consistently throughout. I've found that I've had a much easier time reading other people's code when I've known what kind of format to expect; I've always dreaded debugging code that several people have worked on (something I did a lot of in summer jobs as a student; projects given to students tended to be multi-generational and when I got to it, the mish-mash of different styles made it really hard to understand what was supposed to be going on).

Keep to the same format and conventions throughout and you'll be doing yourself and anybody that follows you a favor.

In keeping in this vein, I don't believe that collaboration on software is necessarily a good thing; many of the world's best and trend-setting software was developed by extremely small teams (less than five people) working extremely tightly and focused. The most monstrous and conservative code is always written by large groups of people.

Social scientists could probably talk about why this happens (i.e., the lack of ownership in big projects, it being very hard for a single voice to be heard, etc.), but the important thing to remember is that it does happen. I also think the ownership issue is important. With small teams, screw-ups can't be blamed on institutional and systemic problems; the source can be identified and the problem can be fixed quickly.

PICs, at most, can only hold a few thousand instructions, so just let one person be responsible for developing the code for the application.

I also want to discuss code-version control. As I've noted, this is a $10 phrase for how you make sure you're using the right source for your application. There are a lot of tools out there that will help you manage the code. To keep it simple, I find that simply putting all the code (except for common libraries) for an application in a single subdirectory to be adequate for pretty much all the PIC coding that I do. This means I can find everything I have written easily and efficiently.

Now for a few words about label and variable names. I try to keep to a few conventions. With labels (either for "gotos" or routines), make them simply descriptive; "flashLED" is a lot better than "flashLED_1sec" (especially if the comments following it explain what is going on). As to variable names, if I require counters or simple indices, I fall back on my FORTRAN programming experience and use "i," "j," "k," and "n." By doing this, no comments are necessary. For other variable names, I try to keep them simple and descriptive for their function. This might even mean that I use two variables when only one is required. You're probably shaking your head and saying that there are only a few RAM registers in the various PICs and I'm squandering them, especially with what I've said about them being a scant resource. Well, if I have the choice between making a program readable or use the fewest possible resources (even if they're in short supply), I'll push towards making the program readable.

Another convention that you will see a lot of in the example programs in this book will be temporary registers. I often use "_1" and "_2" for saving values during a series of operations. Like the counter convention, this one will allow me a couple of registers for multiple purposes and not negatively affect the readability of my code. I also use "_w" and "_status" for saving the "w" and STATUS registers, respectively, in interrupt handlers. These two registers cannot be used for any other purpose because interrupts can happen at unexpected times.

When developing an application in a series of steps, I start off with a simple program that shows me that I've wired the PIC into the application correctly (i.e., just blink an LED). Following this, I work on the code, checking out all the features until I am comfortable the hardware does what I want it to. With this method, I build up to the final application comfortably and only change one thing at a time. This simplifies debugging, because I can focus in on the most likely problem areas for each step of the project.

In terms of labels, there are cases where you don't have to use labels for "gotos," and this is when you know where you are going to be relative to the "goto" or jumping to an absolute address. The "$" character in MPASM is replaced with the current address at assembly time:

```
Finished              ;  Return here when the program is complete
   goto    $          ;  Loop forever
```

I think this is a reasonable use of "$." Deciding when to use "$" instead of a label can be tricky; the criteria that I use is asking myself if using a label will make the code more confusing or more difficult to write:

```
btfss     STATUS, Z        ;  If the Zero flag is set, inc 16-bit variable
   goto     Skip_This
   incf     Reg
   btfsc    STATUS, Z
     incf     Reg + 1
Skip_This
```

can be improved to:

```
btfss     STATUS, Z        ;  If the Zero flag is set, inc 16-bit variable
   goto     $+4            ;  else jump over the increment
   incf     Reg
   btfsc    STATUS, Z
     incf Reg + 1
```

The second example produces exactly the same code as the first, but it eliminates the need for coming up with new and meaningful label names every time you use it. By indenting the code executed when the condition is not true, you are giving the appearance that the code is executed only when the condition is true. This is a nice visual cue that will help you, especially if you are used to working with structured languages.

I recommend using the "$" with offsets in macros because MPASM macros do not produce their own uniquely identified labels. This means that, if you use a label within a macro, it will be repeated each time the macro is used. Now, MPASM knows which ones are used within the macro, but I find it confusing when I'm reading through the listing.

Having said all this, I'm not trying to say that you shouldn't use meaningful label names. I'm just saying that, sometimes it could be appropriate to forgo the label and use an address relative to the current address.

Labels within subroutines are something that can cause confusion. Often, the most appropriate label to use is something like "Loop" or "Skip" or "End." These obviously can't be used repeatedly in subroutines because MPASM will become confused on which one to use.

To avoid this problem, I put the acronym of the routine name in front of each label used within it. The advantages of this are two fold. First, it allows you to keep track of where you are in the routine and notice when you have left it (i.e., by missing the "return" instruction). The second is that it allows me to use basic labels, without worry that I have used them repeatedly. This can be seen throughout the example code provided in this book.

At the end of each development session, the source code and any batch files needed to compile/assemble/link the code are saved on diskette and put in a safe place. Yes, I know it's old-fashioned and anal retentive, but it works and I haven't lost any code in years. By having the code on a separate disk, I can always go back and look at what I had before.

As I've noted elsewhere in the book, interrupts will make your life vastly easier. Having said that, I should discuss the philosophy I use to develop interrupt handlers. As you go through my code, you'll see that I use interrupt handlers as *device* handlers and take the interface from the mainline of the code and put it into the interrupt handler. This means that the mainline is primarily responsible for gross execution control.

Looking back over the previous paragraph, I realized that I should have stated right at the top that the mainline is simply used for gross decision making. This means, that the information provided to it from the various subroutines and the interrupt handler is simply handled by the mainline. Formatting your code in this manner will make your debugging significantly easier, with problems easy to identify, find, and resolve.

Just for something to think about, often in calculations you'll find reasons to use the final destination as an intermediate value. This means that, in coming up with the final result, you will store intermediate results in the register where you are placing your result. In the PIC, this could very easily be an input/output register.

For example, say you wanted to combine two registers together to get an output for a peripheral device:

```
c = a | ( b & 0x077 )
```

In producing "c," you might use the code:

```
movf  a,w                    ;  Get the first value
movwf c                      ;  Store it for later
```

```
movf  b,w                    ;  Get the second value
andlw 0x077                  ;  Clear some specific bits
iorwf c, reg                 ;  Save the result
```

This code will produce the desired end product in "c," but will it do it in the proper manner for the hardware? If "c" is an I/O port or a hardware register and one of the bits in "c" is a clock or a status driver, the peripheral hardware might not work correctly. You might want to change the code to:

```
movf  b,w                    ;  Get the second value
andlw 0x077                  ;  Clear the specific bits
iorwf a,w                    ;  Add the first value to the second
movwf c                      ;  Store the result
```

The second example will make sure that the result is put into "c" at the same time. This is not to say that the first chunk of code is not correct; it might be correct for the device you are interfacing to. I just wanted to illustrate that there are different ways of doing things that will ensure you will not have any problems with peripheral hardware.

As noted elsewhere, there are registers that exist in a bank other than zero. To identify these registers, Microchip has provided bits above the standard 7 (or 8 in the high-end PICs) of the register address. These bits are used to indicate which register page the register is located in.

For example, the TRISB register is defined at address 6 in bank 1. To indicate this, the nominal address of 0x086 is given.

If an address greater than 0x07F is used, the Assembler will return a message "302." To eliminate the warning message, I AND the constant with 0x07F. This places the register within a valid page address range and still keeps the label, which makes following the program easier to understand.

Of course, before ANDing the label value with 0x07F, I make sure that the RPx bits in the STATUS register are set correctly for the bank the register is in. Yes, the message could be suppressed by using the statement "errorlevel 0, –302" (and eliminating the need for the "& 0x07F" at the end of each statement), but I leave the message enabled to tell me where to put the changes in RP0.

One important style of hardware programming that you might want to use is the idea of *device drivers*. Device drivers are blocks of code that control the access to a specific piece of hardware. In the PIC, this hardware could be the I/O ports, the EEPROM memory, serial I/O, or any other separate functional hardware block. The advantages of using device drivers to access the hardware are code consistency (the same method of access is used each time) and reusability (the same hardware interfaces can be used for a variety of different programs). Device drivers also force you to ensure that the hardware that is accessed from a number of different places in the code is always accessed in the same manner (for consistency). Device drivers are probably best known as methods of providing consistent access to types of devices (disk drives, screens, CD-ROMs, etc.) in your PC, but they can be a very useful method of accessing hardware in PICs.

When developing code, don't leave something feeling that it's "good enough." There's no such thing as perfect code, and in reviewing it, you will invariably fix as-yet-unencountered problems or find areas that can be improved upon.

Lastly, when you're developing code, I find it best to always take notes. This way, when you're still trying to validate the hardware and you come up with a more-efficient way of controlling it, you don't have to break off and enter it into your (half-finished) code. Notes are invaluable for when you are debugging and encounter problems you have solved before or have to change various parameters to understand the problem.

As I said in the introduction to this chapter, the whole point of this is to help you develop code that you can easily read and understand. This really isn't an issue when you are first developing the application; however, it is a very big issue when you have left the application for some time and have to fix a problem or add a feature.

One of the sure facts I have learned in this life is, "If you wrote it, you own it." This is true even if you are old, gray, and promoted (as true as death and taxes). The only way you'll every escape your old code is death. So, when you write code, write it in a way that is comfortable for you. Years later, you will be able to look at it and figure out what you were trying to do. What I am trying to say is, as an individual, you have unique thought processes that make some coding styles more efficient for you. Don't try to mask them or hide them or write with somebody else's style; stay true to yourself and you will develop code efficiently and can pick it up years later when a problem arises or modifications have to be made.

Debugging your program

So, now you've written a program in a style you're comfortable with, you've reviewed it a couple of times and worked it down to improve its efficiency, you've programmed it into a PIC, and apply power and . . . .

It doesn't work.

So what do you do?

For the PIC, I have a three-step process for debugging code. It is:

1. Simulate.
2. Simulate.
3. Simulate.

The PIC, like most other microcontrollers, has simulators for use by the developer. You should become very familiar with them. This section will probably touch on a bit of the previous section in terms of style, but the advice should help you avoid getting into situations where all you can do is scratch your head.

I should point out that this section is about software debugging and assumes that your hardware is rock solid. Actually, for the PIC (as is for most other microcontrollers), this is not an issue, as long as you know what the rules are (which are discussed throughout this book and summarized in appendix A).

So what is a simulator? A simulator is a piece of software that runs your code in a "virtual" PIC that you can control the input to and monitor its progress. This is different from an emulator, which actually replaces the PIC with hardware that behaves exactly like the PIC, but who's execution can be monitored.

If the language you are writing in cannot interface with a simulator, then find another language.

"Wow," you must be thinking, "a simulator must be a really wonderful piece of code." No, they are usually terrible in terms of user friendliness, they are not all that accurate, and they can be pretty slow. However, they do give you an idea of what's going on with your application.

The simulator will be your primary window into understanding what you've told your code to do (as opposed to what you want it to do). I have found that there is really no point in trying to program a device until I've simulated it six ways to Sunday and thoroughly understand how the code is going to work.

In chapter 8, I will show you how to use MPSIM to run basic functions and create custom debugging windows and input stimulus. What I want to do here is talk about what a simulator will do for your code development.

Without using a simulator, you really won't know how your application is running. I have been developing code for years on a PC and have always used a symbolic debugger for understanding how the program works and finding problems. The simulator is analogous to the symbolic debugger in that it allows you to understand what is happening.

The simulator also allows you to see what kind of margins you have in your programs. As I noted earlier, a simulator is generally not that accurate. I find that I can never be sure of the timings with a granularity of less than three or four instructions. While you're probably thinking this is a problem, I found that it is actually an advantage; it means that, when I develop my code, I leave in margins sufficient to work around these problems. In doing this, I found that I have reduced the chances for intermittent software errors.

The simulator will help you understand what your margins are and allow you to make sure that they are never violated. I've found that, with as few as five instruction cycles of margin, my code is rock-solid.

In chapters 8 and 9, you will see some places where I have put in conditional assembly to change loop counters and delays. As noted previously, this will reduce debugging time and allow you to get to the critical code quickly without loosing the thread of the program flow.

Make sure you put breakpoints at code addresses where you expect the program to branch to follow the change in input. This will eliminate a lot of the tedium of debugging your code.

Along with using the code, I find it critical to develop proper stimulus files to test my code. Stimulus files allow you to create an input waveform that will be repeated each time you run your program and can be used to debug it.

So, what happens when you've finished your program, loaded it into the PIC, and it runs *most of the time*. You've got the dreaded intermittent errors (which are the worst type). What do you do?

First off, you work at understanding what the input conditions are and the area of the program which might have caused the problems. This can be quite a tedious undertaking. You can create code and interface hardware that outputs critical parameters and variables. (You could even send these parameters to a PC using an RS-232 interface.) You can load the simulator with a number of different variables and working them through. I have carried out all of these actions to try to understand why a problem has been encountered.

Once you understand what the input conditions are that caused the problems, your best plan of action is. . . .

Simulate!

Chances are that one of two things will happen. The first is that the code on the simulator will fail. If this happens, then it should be a simple exercise to understand the problems and debug the code. If it doesn't, then you know the code is good, and you have to figure out whether it is a hardware problem or the simulator. The simulator can't always accurately simulate what's happening when hardware is attached to a PIC (and your stimulus file might not be accurately simulating how the hardware is performing).

However, most of the time your problems are caused by software.

My philosophy for debugging is that it is an exercise in eliminating variables. Before debugging a problem, I find it useful to list out what all the problems could be. With this list, I then develop a list of tests that will tell me whether or not each potential problem is the actual one.

Once this is done, you'll find that debugging an application is not an onerous job that you never succeed at, but is one that quickly becomes part of the development process.

In closing

Looking back over this chapter, I've really dumped a lot of stuff on you. I will try to illustrate a lot of my points in chapter 8, but I know that I won't be able to address everything I've touched on here.

The main points I wanted to cover in this chapter are:

- Develop code that you can write and understand easily.
- Use tools that you are comfortable with.
- Try to write code that is self-explanatory and doesn't require voluminous comments.
- Using a simulator, understand as fully as possible what your code is doing before you burn a PIC and see what happens with the hardware.

Programming PICs

The programmer presented in this chapter was going to be originally a section in chapter 9; however, as I worked on this, I realized that PIC programming is a topic worthy of its own chapter. As I developed my programmer, I realized that there were a number of things that I wanted to say about programming and development systems.

Programming

One of the most powerful features of the mid-range PICs is the ease in which they can be programmed. The programming protocol requires only three pins for the programmer to control. However, there are a number of rules to follow. Understanding how PICs are programmed could affect how your final product is designed and built; taking advantage of PICs with ISP capability can reduce the manufacturing time and cost required to bring a PIC-based product to market.

Low-end programming

The PIC16C5x devices use a very simple parallel protocol. Once the programming voltage (Vpp) is applied to _MCLR, the PIC is in a mode to allow Control Store to be read from/written to.

The program data are read/written in a parallel form, writing all 12 bits of the instruction at a time. Data is strobed in by pulsing TOCK1 low (as shown Fig. 7-1), and the internal PC counter is updated by pulsing OSC1.

You will find that the low-end devices program much faster than the mid-range devices. There are three reasons for this. The first is the serial load of the data in the mid-range devices, which is going to take a number of bits longer than just loading everything in parallel. The second reason is the speed in which the EPROM cells can be loaded in the low-end (10 µsecs versus 10 msecs for the Mid-Range). Lastly the low-end devices don't have the requirement of having to send a command that the mid-range is required to do.

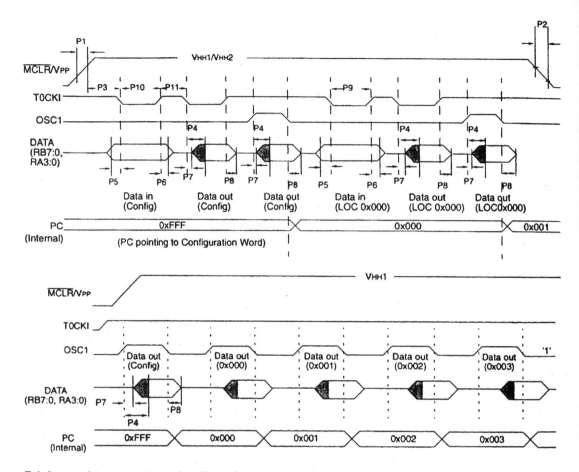

7-1 Low-end programming and verifying the timing waveform.

However, having said this, the low-end doesn't have the big advantage of the mid-range programming protocol; which is In-System Programming.

Mid-range programming

The mid-range PICs use a synchronous serial protocol, which uses very few pins for programming. This is a tremendous advantage for developing hobbyist programmers and, as noted elsewhere, allows the parts to be programmed *after* they have been soldered into a card. (See Fig. 7-2.)

The EEPROM/Flash parts are programmed slightly differently from the EPROM parts in that they have additional instructions for clearing the EEPROM and programming the EEPROM data memory and do not require an "End Program" instruction after the data has been loaded into the PIC.

The mid-range parts differ from the low- and high-end devices in terms of programming not only because they take the data serially, but also because they require instructions for each programming step.

All this will be shown as I go through the programmer circuit for the book.

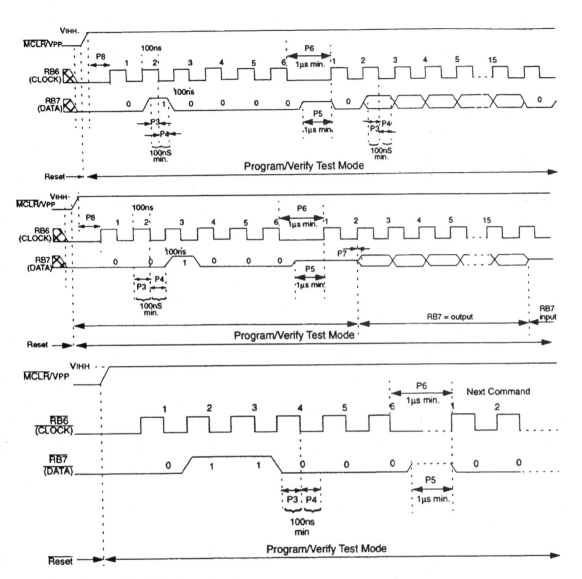

7-2 The mid-range "Load Data" command.

High-end programming

The high-end PICs use a parallel protocol that is similar to the one used by the low-end Devices but with one major difference: The address to be programmed is passed along before the data. (See Fig. 7-3.)

Available programmers

There are literally dozens of PIC programmers available. Maybe I should clarify this before going on too far. There are dozens of 16C84 (and, because of similar fea-

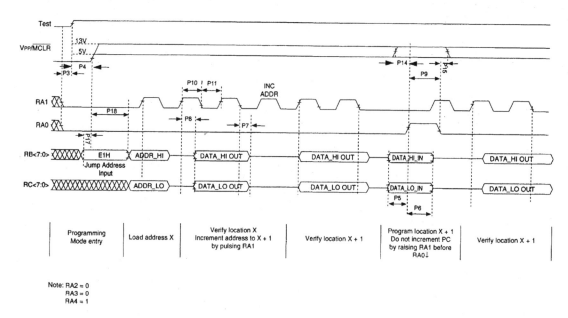

7-3 High-end programming and verifying timings.

tures, the 16F84) programmers available, many mid-range programmers available, and a few low- and high-end programmers available.

For both the low-end and high-end, I would probably recommend buying a programmer designed for them, rather than trying to build one from scratch (as you can do with the mid-range PICs). This is because the parallel protocol requires quite a bit of hardware to make sure everything is being handled correctly.

For this reason, when I am talking about available programmers, I am primarily talking about mid-range programmers (and 16F84 programmers specifically).

They can be unbelievably simple, (See Fig. 7-4 for an example), or they can be production multiple-part programmers costing tens of thousand of dollars.

Before investing in a programmer, I recommend trying to understand what your requirements are. For the hobbyist, a programmer like the one shown in Fig. 7-4 (yes, it actually works! Look for "COM84" on the Web), which can be built for a very modest amount of money, might be all that's required. If you are planning on production, then making sure that the Microchip "Production" specs are met is a requirement, and the programmer might end up being a much more substantial piece of equipment.

It's also important to understand what parts you are going to be programming. The programmer shown in Fig. 7-4 will work fine for a 16F84, which has very modest current requirements for Vpp (measured in microamps), but will not be acceptable for any EPROM-based PICs (which typically require tens of milliamps) or low- or high-end PICs, which use a parallel protocol.

In appendix D, I've listed a number of sources for programmers along with their characteristics. By no means is this a comprehensive list. Before investing a large amount of money on a programmer, research the various ones that are available and find the one that best meets your requirements.

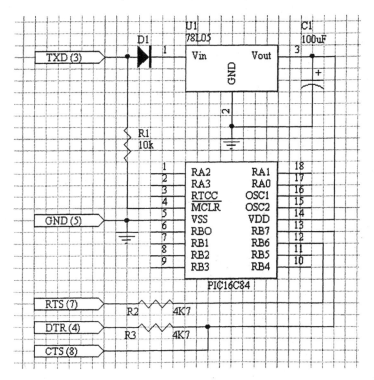

7-4 "COM84" programmer circuit.

Before presenting my programmer, I just wanted to list some questions you should ask yourself before selecting a programmer:

- *Do I want a "production" or "prototype" programmer?* The Microchip specifications on what is a "production" quality programmer are quite specific. (These specifications largely center around how the part is verified.) If you are going to be shipping product with a PIC inside, then you should seriously consider using only a production-quality programmer.
- *What will be my host system?* As I write this, there has been a lot of discussion on the PICLIST about what systems programmers, emulators, and the Basic Stamp should run on. If you have an IBM-compatible PC running DOS or Windows (up to Windows '95), you won't have any problem. If you aren't, then you should look very closely on how the programmer will interface with your hardware and software (both are an issue).
- *Do I want an integrated development system?* For most of my development, I use a PICStart Plus, which is integrated with MPLAB. If I were to use another programmer, I wouldn't get the advantages of downloading the code I'm editing, assembling and simulating.
- *Am I going to program just 16F84s or EPROM parts as well?* As I've mentioned elsewhere, the EEPROM/Flash (16F84) parts require Vpp currents on the order of µAs, while EPROM parts (the rest of the mid-range line) require tens of mAs. The simple programmer circuit presented in Fig. 7-4 will not be able to program EPROM parts.

There are no right and wrong answers to these questions. However, before buying or building a programmer, you should ask yourself these questions to ensure that what you get meets your needs.

PROG39: The YAP

As one of the projects in this book, I have created a mid-range PIC programmer with the capability of being put into a prototype circuit and providing a PIC with ISP capabilities. (See Fig. 7-5.) Originally, the aspect of programming PICs was a sideline; the basic purpose of this project was to show how the PIC could communicate with a PC in a fashion other than RS-232 (as a number of the projects communicate using the Serial Communication Ports of a PC). As this project progressed, I realized that using the In-System Programming capabilities would allow me to simplify the work done in chapter 8 and allow testing out applications with the programmer as part of the circuit.

As I originally designed the circuit, it would connect to an IBM-compatible PC running a PC-DOS application. The PC would provide control over the programmer as well as provide timing information for the programming of the PIC.

The initial programmer worked quite well—but only on the PC that the code was developed on. This lead me to scramble a bit and try to figure out how to provide consistent delays across a reasonable range of PCs (what I found was that it's just about impossible to create consistent timings using the wide range of PCs out there). At the same time there was a lot of discussion on the PICLIST about how some of the very simple programmers (like COM84 in Fig. 7-4 above) wouldn't work on different

7-5 The "YAP" programmer in a protoboard.

PCs. Along with this, there were complaints by Mac, Linux, and workstation afi-cionados complaining that there weren't programmers available for their systems.

This all happened while I was working on the MIC and thought about developing a programmer that would take downloaded information and put it into a PIC without any processing from the PC (now known as the "host") other than what was required to download the .hex file from the terminal emulator.

This re-evaluation made me realize two things: such a programmer would be us-able from a wide variety of systems (basically anything with an RS-232 port and a terminal-emulation program) and the programmer's PIC code would be a lot more complex. However, the increase in functionality (and not having to worry about whether or not the programmer would work on different PCs) would really make it all worthwhile.

With this, I then created a new set of specs for the programmer:

1. Able to program all mid-range parts (EPROM and EEPROM/Flash).
2. Able to put the programmer into a prototype circuit as an "ISP" part.
   2.1 As part of "2," design in enough extra power capability to eliminate the need for a separate power source in the prototype circuit.
3. Allow any RS-232 equipped host to communicate with the programmer.
4. Allow serial communications between the prototype circuit and the host as a method of passing debug messages and user instructions.

The last item in the list was put in because I had just finished the serial-LCD in-terface and I thought that this would go one step further than hanging the SLI onto the prototype circuit.

Because of the complexity of the programmer, I would like to go through the major functional blocks of the hardware and then explain how they all work to-gether.

The power subsystem (Fig. 7-6) is actually a simple dc power source with two switchable supplies. Overall power is provided by a "wall wart" (an inexpensive wall-mounted ac/dc converter, often available at places like Wal-Mart). The "wall wart" should be set to at least 17 V or 18 V output so that the 78L12 circuit will provide a full 13.4 V (the two diodes placed in series with the ground reference will bring the 12-V output to 13.4 V, which is within the recommended range for all PIC devices; 12 V is marginal for some EPROM parts).

The programmed parts (referred to as U2 in this section) Vdd and Vpp are con-trolled by U1 (the controlling PIC), turning on and off two opto-isolators. This allows the PIC to be fully powered down and removed/replaced in the YAP without disturb-ing any other parts of the prototype circuit.

The clock subsystem uses a programmable clock to provide clocks to both U1 and U2. The primary output clock (16 MHz) is fed into U1, while the programmable clock (frequency controlled by U1) is provided to U2 through another switching cir-cuit. The 16-MHz clock input into U1 means that it has to be a part like a 16C61 ca-pable of running at up to 20 MHz. (See Fig. 7-7.)

Q1 is used along with U1 RB3 to enable the clock going to U2's OSC1 Pin. When RB3 is low (during device programming or U2 idle), no clock will be passed to U2. When this line is high, an inverted programmable clock will be provided to U2's OSC1.

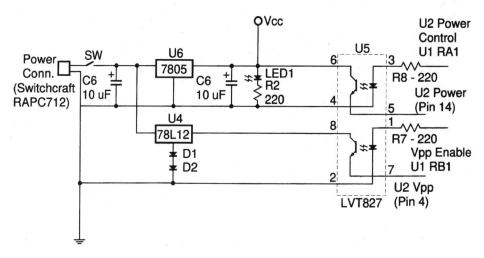

7-6 Programmer power circuitry.

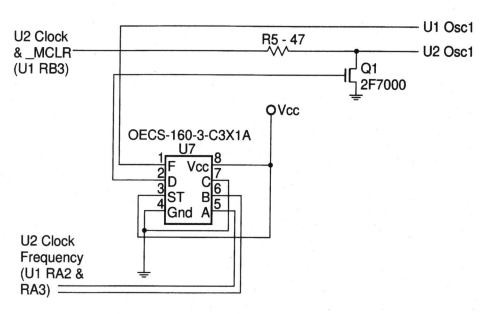

7-7 Programmer clock design.

The last major subsystem is the host/RS-232 communications. (See Fig. 7-8.) Communication with the host is carried out by an RS-232. I use the Dallas Semiconductor "DS275" for the CMOS-to-RS-232 level translation. This device is designed for half-duplex RS-232 communication. During transmission, the DS275 uses the negative voltage from the RS-232 input (pin "Ri") and outputs as a transmit a 1. For the positive 0 voltage, the power supply's 5-V output is used. This circuit has been used

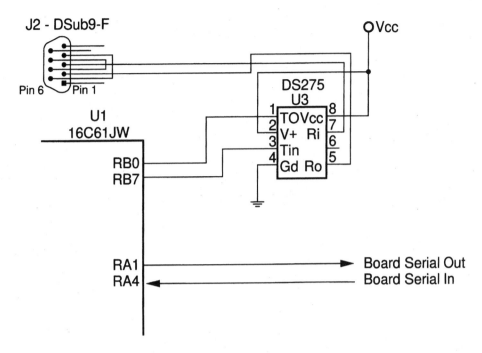

7-8 Programmer communications.

on a variety of host systems without any problems and is much simpler to wire than the "typical" 5-V-to-RS-232 level translator, the MAX232.

At the bottom of the communications schematic, you'll notice that I have put in two lines "Board Serial In" and "Board Serial Out." These are used to allow serial communications with the host while U2 is running.

The communication between the host terminal session and U2 put me in quite a quandary, because I would be using U1 to pass data between the host and U2 while "eavesdropping" on what the host was sending (waiting for a Ctrl–C or 0x003 character to end U2's execution).

This was resolved by using a TMR0-based interrupt at three times the serial data speed (1200 bps) and polling each of the incoming lines and passing the data along to its destination. Actually, this method of reading the incoming serial line was used for all serial communications (rather than my "typical" method of interrupting the PIC when the Start bit comes in). The code to do this is quite an efficient method of providing full-duplex serial communication. The only problem I have with it is that it executes continually, and if you have a period of no serial activity, the PIC cannot stop executing this code in case a data packet is sent to the PIC.

With the subsystems I've described, it should be obvious that the U1 circuit (or subsystem) is really just a connection to each of the other subsystems with a 4.7K resistor pull up to _MCLR. U2 (the programmed PIC) is in a similar situation, either being connected to the subsystems described previously and/or the 19-pin prototype circuit connector.

This connector has the pin out shown in Table 7-1.

**Table 7-1. The pin out of
the 19-pin prototype
circuit connector**

Pin	Description
1	Ground
2	Vcc
3	U2 _MCLR (U1 RB6)
4	U2 OSC2
5	Serial In (U1 RA4)
6	Serial Out (U1 RA1)
7	U2 RA0
8	U2 RA1
9	U2 RA2
10	U2 RA3
11	U2 RA4
12	U2 RB0
13	U2 RB1
14	U2 RB2
15	U2 RB3
16	U2 RB4
17	U2 RB5
18	U2 RB6
19	U2 RB7

With the subsystems defined and the Prototype circuit connector pinout defined, I came up with an embedded card for the YAP. (See Fig. 7-9.), which have the parts installed at the locations shown in Fig. 7-10.

The U2 socket is designed for a ZIF socket. I have built YAPs with both a regular DIP socket and ZIF socket without any problems.

I should point out a couple of things on the RS-232 interface. The 9-pin female D-shell connector is wired as a "DCE" (Data Communications Equipment) RS-232 standard. To connect the YAP to a host, a straight through (not null-modem) cable with a male RS-232 connector should be used. You should notice that the DSR-DTR and CTS-RTS handshaking lines are shorted together. This was done to ensure that the host terminal emulator would work if hardware handshaking was enabled.

The user interface is very simple, with only five different types of instructions. The programmer communicates with the host at 1200 bps. (Yes, I know that this is slow but I'll explain why later.)

When you first power up the YAP, you'll notice the Prompt:

```
PIC Programmer
F> _
```

As I've mentioned previously, this programmer can program both EPROM or EEPROM/Flash PICs. The "F" indicates that the YAP will program Flash/EEPROM parts; an "E" indicates that the programmer is configured for an EPROM part.

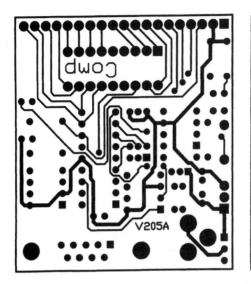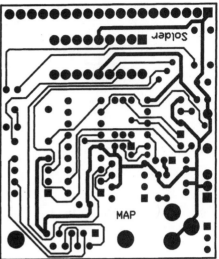

7-9 Programmer raw card design.

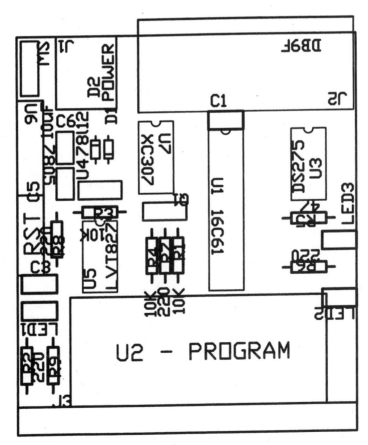

7-10 Programmer component layout.

The complete list of YAP instructions are given in Table 7-2.

Table 7-2. The YAP instructions

Instruction	Description
E|F	Select PIC program store type
P	Begin programming the part
V	Verify the contents of the part
B	Blank check the EPROM part
1|2|4|8	Run the PIC at the specified frequency (in MHz)

Running the PIC in the U2 frequency socket has pretty much been explained earlier. When the number indicating the clock is entered, U1 sets the U2 clock frequency then powers up U1 (and turns off Reset at the same time). While U2 is running, U1 watches the serial data coming in and turns off U1 when a Ctrl–C is encountered. You should notice that there is a "Reset" push button on the YAP. Pressing this activates reset and pulls down U2's _MCLR.

Warning: If you are going to remove the PIC and there are devices that are driving a voltage into the PIC, it is highly recommended that the power be turned off (and these devices stop driving the PIC), because having unbalanced voltages on the PIC could damage the PIC and/or the circuits that are both driving and receiving the PIC lines. U1 is not at an issue here because the PIC programming lines are set to Input (not driving) when U2 is not being programmed.

The programming operations follow the specifications given in the Microchip mid-range PIC programming datasheet. Vpp is enabled with Vdd to put the PIC into programming mode. Once the PIC is in programming mode, U1 either does a blank check (for EPROM Parts) or erases all the EEPROM/Flash memory. Once the blank check/erase is complete, you will be prompted to start downloading the .hex file.

The .hex file should be downloaded as a text file (which imbeds both CR and LF). While nothing appears to be happening during the download, the data is being transferred and the PIC is being programmed.

The program is loaded into the PIC by downloading the .hex file produced by MPASM and the other compilers.

If you were to print out a .hex file, each line would look something like:

```
:0400000008C72020ED
```

In this example, the Line can be broken up as shown in Table 7-3.

To program a part, I use this hex file information to provide the YAP with both the data and the programming timing delay as it is downloaded from the host.

For each line, the programmer carries out the following algorithm:

```
if Byte.0 != ':'                          ; Invalid .hex file format
  goto Error
LineCount = ( Byte.1 + Byte.2 ) / 2     ; Get the number of instructions
                                          ; on the line
 Addr = ( Byte.3 + Byte.4 + Byte.5 + Byte.6 ) / 2
```

```
    if Addr > ( PrevAddr + 50 )             ; Can we get to the address?
      if Addr >= 0x02000                    ; Configuration area?
        goto Error                          ; No, too big a jump
      else
        Send LoadConfig                     ; Yes, move PC to config area
    else
      while Addr != PrevAddr                ; Inc to the expected address
        inc PC                              ; Inc the PICs internal PC
        PrevAddr = PrevAddr + 1
LineLoop                                    ; Handle each instr from here
  Wait.for.four.characters.to.come.in       ; wait for the data
  Instruction = Next.Four.Characters        ; Get the instruction
  program( Instruction )                    ; Program the PIC
  inc PC                                    ; Keep track of where PC is
  Addr = Addr + 1
  LineCount = LineCount - 1                 ; Dec the instruction count
  If LineCount != 0
    goto LineLoop                           ; If not 0, do next instr
  PrevAddr = Addr                           ; Save last addr
```

Table 7-3. The breakdown of a .hex file line

Byte	Description
0	Always ":"
1–2	The number of instructions × 2 to load
3–6	The address the instructions start at × 2
7–8	"00" for data, "01" for .hex file end
9–12	The instruction in Intel hex format (low byte first)
.	More instructions (up to 8 on a line)
:	
n − n+1	The file checksum
n+2 − n+3	CR/LF

This algorithm uses the time for each of the 4 data bytes of the next instruction to provide a delay for the PIC to program in the previous instruction.

Ideally, I would have liked to provide the YAP with a 9600 bps interface; however, at this speed, four instructions take a minimum of 4.2 msecs to be transmitted. This is too short for the program operation to complete (the specification is for 10 msec). Yes, 2400 would also work, but I felt that 1200 bps was more of a standard speed.

At 1200 bps, it will take just under a minute to load a 1K instruction program into a PIC.

If there is a failure of any type, the YAP waits for the download to end and then reports on the error.

One type of error that might be surprising is a "Spacing Error." As can be seen in the code, if the specified address is different from the next address (stored in the PIC's PC) by more than 50, then the programming is stopped. This is because the

Program Counter incrementing is done when the data indicator (the "00" at bytes 7 and 8 of the line) and the first four bytes of the instruction are being received. During the time needed to spend these bytes, no more than 50 "PC Increment" instructions can be sent to the PIC.

This is not an issue with the configuration area because there is a special "Configuration Load" command that completely bypasses the standard process and changes the Program Counter to 0x02000.

If you read through Microchip's mid-range programming data sheet, you'll see that EPROM parts can have each individual instruction EPROM hit with a instruction and miscompare up to 25 times before there is an error. To allow some measure of hitting EPROM again, you can initiate an EPROM write right after the programming has failed, and the "Blank Check" will not run again. Actually, this can be useful in patching EPROM contents (as described elsewhere) without having to erase the EPROM and reprogram it.

An important feature of this programmer is its programming of the configuration fuses directly using the values in the .hex file. There is no facility to manually set the configuration fuses. The reason for doing this is to make sure that no bits get inadvertently and incorrectly set during programming. This is a very big consideration with any PIC (because setting the wrong type of oscillator or enabling the WDT when you don't want it enabled can lead to a PIC that doesn't seem to be running) and is exceptionally important with the new EPROM parts that have U/V protected code protect bits (which means they have been designed so that they cannot be erased). This means that a __CONFIG statement has to be put in all your programs (which I feel is good practice anyway).

There are no plans to market the YAP. This is simply because there are so many PIC programmers available (this is where the name comes from: "Yet Another Programmer"). Not to mention the fact that, to build this programmer, you need a programmer in the first place (to program U1).

However, if you would like to build the YAP, please let me know, and I'll help you out with getting raw cards and programmed parts.

8
CHAPTER

Experiments

In this chapter, I would like to go through how the PIC works, both in terms of hardware and software. While they are written with the software tools provided with this book (MPASM and MPSIM) and the programmer/development system outlined earlier in the book, the experiments presented here can be done on a variety of different platforms, using different development tools. All the experiments can be set up in a matter of minutes and do not require any special tools or parts.

Tools

To implement the programs and applications given in this chapter, the following tools are required:
- 16F84 programmer
- IBM-compatible PC with Microchip's MPASM and MPSIM installed
- Protoboard (with jumpers)
- 5-V power supply
- (Digital) multimeter

Some optional tools that will make the work easier include:
- Logic probe (highly recommended)
- Oscilloscope (minimum 20 MHz)
- Logic analyzer
- PIC emulator

The following parts will be required for the experiments presented in this chapter:
- PIC16F84/P-04 (4 MHz, plastic 18-pin DIP 16F84)
- 0.1 µF decoupling cap
- 1-MHz crystal or ceramic resonator
- 2x 15 pF capacitors (if a crystal or two leaded resonator is used)
- 10K 0.25-W resistors
- Several LEDs with an equal number of 200-Ω to 270-Ω resistors

- A few momentary-on push-button switches
- Miscellaneous resistors
- 10K Pot
- 74LS244s and 74LS374s
- An 8-switch DIP switch with 4.7K 10-pin resistor SIP

I want to go through each of the tools individually to explain what they are used for and any requirements there might be for their use. The tools and parts I have chosen are either available in a typical hobbyist's home or can be bought very cheaply.

In this book, I have provided a programmer/development system (see chapter 7) that can be used for almost all the experiments in this chapter. The advantage of this programmer is that it does not require separate clock, reset, and power circuitry. As I have said, there are literally dozens of different 16F84 programmers out there, and any one would be fine for use with these experiments. Different hobbyist 16F84 programmers, along with software, can be found at the various Web sites listed in appendix D.

One of the little secrets I've discovered over the years is that development kits give you more than you pay for. The Microchip PICStart Lite and PICStart Plus kits live up to this rule. The packages include a programmer; Microchip databooks and application guides; documentation for the included assembler, simulator, and programmer; and sample components. If you buy a PICStart B1 kit, you should be aware that the assembler shipped with it is a downlevel version called MPALC. This should be upgraded to MPASM from the Microchip BBS, ftp site, or Web page. The PICStart Plus development kit can be used seamlessly with MPLAB and can be used to program the entire range of PIC devices. These programmers, while being more costly than some of the hobbyist programmers available, will give you everything required to start developing PIC applications.

For the projects presented in this book, the choice of programmer is a bit more critical because PICs other than the 16F84 are used. Along with a programmer capable of programming a wide range of PICs, you should buy a UV EPROM eraser. Hobbyist erasers can be found quite cheaply from the sources listed in appendix D.

The development system (PC) should be the equivalent of at least an IBM-compatible PC with a 386 processor running PC-DOS 5.0 or better. There are PC emulators and development tools written for different platforms (e.g., the Apple Macintosh), and these can be used as well, as long as they can run MPASM or assemble MPASM source code. For simulating the programs, files for the Microchip MPSIM program are provided.

As I go through the experiments, I will reference my programmer along with MPASM and MPSIM.

Building the applications can be done in any way you desire. I used a protoboard (known under many different trade names) for the examples used in this chapter because it is reasonably cheap and easily changeable. Figure 8-1 shows the protoboard with the programmer.

The protoboard consists of an array of holes 0.1" apart, in which wires or component leads can be pushed into the boards. The connector edge on the programmer

8-1 "YAP" programmer and protoboard.

has pins 0.1" apart and can accept wires of approximately 22 gauge. The Vcc and Gnd signals coming from the programmer should be wired to the long busses on the protoboard to provide power for the external devices.

If my programmer is not used, then the protoboard should have a 5-V power supply for Vcc (Vdd) and Gnd, and you should attach the lines to the busses on the protoboard. The actual supply should be capable of sourcing 200+ milliamps to power both the PIC and external circuitry. A good idea is to always put an LED (and a current-limiting resistor) on the board that will light when power is applied to the board. This setup will be useful when experimenting further with the PIC, either your own designs or trying out the projects presented in this book.

One thing I haven't suggested yet is using a ZIF (Zero Insertion Force) socket for the PIC when developing/debugging your circuits. This little tool can be invaluable in allowing you to remove the PIC over and over again without damaging the component's leads. 18-pin ZIFs can be hard to find (I use the Textool 18-pin Zip Dip Socket), but they will make your life a whole lot easier when developing applications.

A simple multimeter will allow you to check over your circuit. Make sure that you have one with voltage, current, resistance checking, and audible circuit (diode) checks. Capacitance measurements are something nice to have as well.

A logic probe, while not absolutely necessary, will be very useful to you as well. Along with indicating logic levels in your circuit, a probe can be used to determine whether or not a port is driving and whether or not your oscillator is running. They are generally quite cheap. If you do buy one, it will be $20 very well spent.

For the well-heeled, an oscilloscope will be of great value in debugging your circuits and seeing what is going on in them. Ideally, a digitizing (storage) scope with a bandwidth of at least 25 MHz would be best, in order for you to capture one-time-only events. While a scope isn't absolutely required, it will make your debugging a lot easier.

A logic analyzer will give you the ability to watch signals directly on the board. It's an excellent tool for debugging and "learning" the circuits your applications will be interfacing to. Like the oscilloscope, there are PC adapters that can be used as logic analyzers. While I feel that a logic analyzer is absolutely required for highly complex (with high parts count) projects, it really isn't required for the applications presented in this book.

An emulator is essentially a PIC chip that is interfaced directly by control hardware. The "PIC Master" emulator is very expensive ($2500) but very necessary for high-end applications, with very critical timing and software-reliability requirements.

Since developing the emulator presented in the book, I've found it invaluable for debugging applications (the test case was the servo controller, and I was amazed at the number of noncritical bugs that I found in the code). However, it is not a project you should begin without going through this chapter at least.

The tools recommended here will allow you to replicate the examples and applications contained within this book as well as develop a few of your own.

For the rest of this chapter, I would like to go through the various features of the PIC 16F84 and show what the device can do as well as discuss some trapfalls you might encounter.

You will note that I will make use of the simulator quite a bit instead of actually programming and running hardware. This is because a lot of the problems that you can observe easily in the simulator simply don't make sense when hardware is involved. The PIC doesn't naturally have interfaces for you to use to understand what is happening, which makes simulation of the software very important. Often if a program malfunctions, the PIC will just sit there, and you'll have no idea what is happening. So, I highly recommend you go through every program listed here in the simulator before burning it into a PIC.

All the programs have been tested and will work if the hardware is assembled correctly. I encourage you to experiment with various aspects of the circuits. There are two things to remember when you are changing the circuit:

- Only change one thing at a time. If you have an idea to improve or change several aspects of the circuit, do it in stages. If a change results in the application not working, then the problem should be easy to find based on the last change. Changing several variables and then retesting will make application debug just about impossible.
- Always remember that if you get lost and the circuit no longer works, go right back to the previous step and begin the modification again (remembering the rule above).

Experimenting with the PIC

Before we start loading programs into an actual PIC and trying programs out in hardware, I first wanted to go through a number of experiments showing how the PIC CPU and software works. The purpose of these experiments is to give you some experience with the PIC and give you some confidence before you start burning code into hardware and getting frustrated because nothing works.

As I go through the experiments, I will start by copying in the entire program and then just putting in the most important aspects of the experiment.

Loading the files

The diskette given with this book contains the current (as of the book release) MPASM PIC assembler and MPSIM PIC simulator (including the PIC .INC files) and all the files used for all the programs presented in this book (Assembler source, "MP-SIM.INI," and stimulus source files).

Follow the instructions in the "README.TXT" file contained in the root directory of the diskette for loading the files onto the hard drive of your PC. Remember to modify the PATH statement of your PC's "AUTOEXEC.BAT" file to pick up MPASM and MPSIM from the PIC/MPASM subdirectories when you are assembling/simulating from other subdirectories.

Later versions of MPASM and MPSIM (along with MPLAB) can be downloaded from the Microchip web site:

```
http://www.microchip.com/
```

PROG1: Register addressing

The first program is a simple register addressing example and runs through the three different ways of manipulating data in the PIC processor. The program is shown in Fig. 8-2.

This program, like all the others in the book, is available on the disk included with the book. To keep you from typing in all this information, I have provided all the source files, appropriate simulator MPSIM.INI files, and any simulator stimulus files that might be required. The files on the disk should be installed on your PC's hard drive as explained in the "ReadMe" file on the diskette.

To assemble and simulate this program, go to the subdirectory containing PROG1 and assemble it using the command:

```
C:>MPASM PROG1
```

A new screen will be displayed; you can watch the source file, along with "P16F84.INC," being assembled.

8-2 PROG1 Simple addressing demo.

```
title   "PROG1 - Simple addressing demo"
;
;  This is a demo of the various addressing modes.
;
;  Myke Predko
;  96.04.20
;
   LIST P=16F84, F=INHX8M, R=DEC
   errorlevel 0,-305
   INCLUDE "PIC\MPASM\p16F84.inc"

;  Registers
Test equ 12                        ;  Just put in one

  __CONFIG _CP_OFF & _WDT_OFF & _RC_OSC

   PAGE
;  Mainline of PROG1

   org       0

   clrf      Test                  ;  Initialize test register

   movf      Test, w               ;  Get the test value
   addlw     'A'                   ;  Turn it into an ASCII
   movwf     Test                  ;  Store the new value

   movlw     Test                  ;  Access test via the FSR register
   movwf     FSR                   ;  Save in the index register

   movf      INDF, w               ;  Get the value from the index
   incf      INDF                  ;  Increment the original value
   movwf     INDF                  ;  Save the unincremented value

Finished                          ;  Now, just spin forever
   goto      $
   end
```

Once assembly is complete, you can press any key and return to the DOS prompt. I have included the line "errorlevel 0,−305" to avoid a "message" indicating that, in the line where "incf INDF" is located, the default destination (back to the register) has been selected. This goes back to the comments made in chapter 6, where I want the program to run without errors, warnings, or messages. I feel that indicating that the destination is back in the register is redundant, and simply putting in a command like "incf Reg" indicates that the register is simply incremented and the results are not placed anywhere else.

Now, you are ready to run the simulator.

With all the programs, I have included an appropriate "MPSIM.INI" file for loading into the simulator. The "MPSIM.INI" file should be located in "*D*:\PIC\MPASM\PROG1" and has been customized for loading "PROG1.ASM" into

MPSIM. When using MPSIM, I always put in a program load command ("lo") as one of the last instructions in MPSIM.INI to save me from having to load the file manually.

To execute MPSIM, simply type in:

```
MPSIM
```

in the subdirectory containing the source and the listing files. It is important to keep these files in the same subdirectory; otherwise, MPSIM will not work properly with the symbolic information. The top "window" on the screen shows you the contents of the various registers in the simulated PIC. The lower window contains an input screen where instructions can be entered and data displayed. The "%" symbol is the command prompt.

In the last line of the "MPSIM.INI" file that I have included with the .ASM file, I have put in the command:

```
DI 0,0
```

This will display the first executable line, which might not be displayed normally during MPSIM's execution.

The first instruction, as expected, is displayed on the last line of the lower window:

```
clrf        Test
```

This line initializes the RAM register (variable) "Test" to zero. Actually, this last sentence should be changed to: This line initializes the RAM register (variable) at address "Test" to zero. Because the PIC has separate data and control memory, variables are not explicitly defined as in a traditional architecture processor.

In a traditional architecture processor, variables have to be explicitly stated and memory allocated because they can affect the addressing of the code. In the PIC, this is not an issue because variables and variable size do not affect the addressing of the code.

Before executing the single-step instruction, you should be sure that the various file registers are all initialized to zero. Assuming this can cause problems later because, in the PIC this is probably not the actual power-on value. Upon power up, all RAM registers contain an undefined value. MPSIM defaults all the file registers to zero upon the start of the program.

The actual power-up RAM register values are investigated later in this chapter.

Use the instruction "ss" to begin single-stepping through the program. "ss" is keyed in at the prompt ("%"), then you press the Enter key. This will initialize the PIC's registers and execute the "clrf" instruction. Once the "ss" instruction has executed, you will see that the hardware registers now display a more familiar value.

The next three instructions add the ASCII value of "A" into the register at address "Test":

```
movf        Test, w
addlw       'A'
movwf       Test
```

The first instruction loads the register at address "Test" (called "Test" from now on for brevity) into "w." Next, the ASCII value for "A" is added to "w." Finally, the result is stored back into "Test." Following these three instructions, you can see that the register at "Test" (address 0x0C) contains the value 0x041, which is the ASCII value of "A" added to zero.

This series of instructions follows the format for a "load-process-store" operation in a traditional architecture processor. In the PIC, the optimal way of carrying out this operation would be:

```
movlw       'A'
addwf       Test, f
```

The addition value is loaded into the "w" register and added to the desired variable, and the result is stored back in the variable. This is one of the reasons why the PIC can run significantly faster than traditional architecture processors.

The next two instructions load the address of "Test" into the index register (FSR):

```
movlw       Test
movwf       FSR
```

After these two instructions have executed, "Test" can be accessed directly (as was done in the first four instructions of the program) or indirectly (as the next three instructions demonstrate).

The following three instructions demonstrate both how indexed addressing works in the PIC as well as the PIC's ability to manipulate data without affecting the contents of the "w" register.

```
movf        INDF, w
incf        INDF                ;   NOTE: Dest of Incf is back in "INDF"
movwf       INDF
```

The first instruction loads the register pointed to by the index into "w." The next instruction increments "Test" and stores it back into "Test" without affecting the contents of "w." Finally, the contents of "w" (the original value of "Test") is put back into "Test," restoring it to its original value.

Now, at the end of the program, I make sure that there is no opportunity for the PIC to begin running amok. I do this by executing an endless loop using the instruction:

```
goto        $
```

As explained elsewhere, the "$" operand actually points to the current address. This instruction is actually telling the PIC to change the Program Counter to the current address rather than incrementing it. By looping in this manner, there is no way that the PIC can ever execute instructions that are not planned for the application.

PROG2: Register bank addressing

The second program to be executed is actually quite a bit simpler than the first. (See Fig. 8-3.)

The purpose of the program is to show how the page register ("RP0" in STATUS) is used to access registers in bank 1 of the register space.

The program can be represented by the following pseudo-code:

```
PORTB = 0                   ;  Clear all the bits for port "B"
TRISB.0 = 0                 ;  RB0 is set to output
Loop Forever
```

8-3 PROG2 Turn on a LED.

```
title   "PROG2 - Turn on a LED"
;
;   This is the first program to be burned in and run in a PIC.
;
;   The program simply sets up bit 0 of port "A" to output and then
;     sets it low.
;
;   Hardware notes:
;     Reset is tied through a 4.7K resistor to Vcc, and PWRT is
;     enabled
;     A 220-ohm resistor and LED are attached to PORTB.0 and Vcc
;
;   Myke Predko
;   96.05.17
;
    LIST P=16F84, R=DEC
    errorlevel 0,-305
    INCLUDE "\PIC\MPASM\p16F84.inc"

;   Registers

    __CONFIG _CP_OFF & _WDT_OFF & _XT_OSC & _PWRTE_ON

    PAGE
;   Mainline of PROG2

    org     0

    clrf    PORTB               ;   Clear all the bits in port "B"

    bsf     STATUS, RP0         ;   Goto bank 1 to set port direction
    bcf     TRISB & 0x07F, 0    ;   Set bit 0 to output
    bcf     STATUS, RP0         ;   Go back to bank

Finished                        ;   Loop forever with the LED on
    goto    $

    end
```

Clearing the port "B" register is accomplished with the "clrf" instruction. The PORTB register is located in bank 0 (which is what the PIC is using upon power-up) while the TRISB register is located in bank 1.

To access TRISB, the "RP0" bit of the STATUS register must be set. This is done by using the bit set instruction ("bsf"). Once this is done, only the registers that are in bank 1 and the shadowed registers from bank 0 can be accessed. Changing the access back to bank 0 is accomplished by simply resetting the "RP0" bit.

Rather than changing the bank-select bit ("RP0"), the "tris" instruction can be used to set the PORTB state.

So, instead of:

```
bsf          STATUS, RP0              ;  Goto bank 1
bcf          TRISB & 0x07F, 0         ;  Make RB0 an output bit
bcf          STATUS, RP0
```

you could use the following two instructions:

```
movlw        0x0FE         ;  RB0 is output/everything else is output
tris         PORTB
```

If you were to change the three lines to the two and assemble them, you would get:

```
"Warning 224: Use of this Instruction is Not Recommended"
```

on the line before the "tris PORTB" instruction.

This is because the "tris" (and "option") instructions are available to allow porting code from the low-end PICs to mid-range with the fewest amount of source code changes. If you are creating code from scratch for the mid-range PICs, use of this instruction is not recommended because it might not be available in future PICs.

Even though PORTB will show all its bits as "0" before TRISB.0 is set to "0," they are actually set to input. To change the input state, you would have to create a stimulus file with all the bits set to "1."

PROG9: The STATUS register

A good understanding of the processor STATUS register flags (Zero, Carry, and Digit Carry) is critical to being able to develop and debug applications. These register bits are primarily used for providing a simple hardware interface to a previous operation. These bits are positive active, which means the condition is true when they are set (equal to "1").

The Zero flag is set when the result of an operation is zero. Actually, it might be more accurate to say that the Zero flag is set when the data leaving the ALU is equal to zero. I have put in this clarification because of the PIC zero register check "movf Reg, f." In this instruction, the contents of the registers are run through the ALU and stored back into the register. (See Fig. 8-4.)

The Carry flag is normally set/reset after an addition/subtraction or rotate instruction. For addition, the Carry flag is set when the result is greater than 0x0FF.

Subtraction, as I've pointed out elsewhere, is another matter because of its operation. Remember that subtraction is actually negated addition:

```
subwf  Reg, w
```

is actually:

$$w = Reg - w$$
$$= Reg + (w \wedge 0x0FF) + 1$$

You can see that the Carry flag is reset when "Reg" is less than "w," but it is set when "Reg" is greater than or equal to "w." This can be reasoned out by noting, if "w" is greater than "Reg," then its negative will be less than "0x0100 − Reg." If this is true, the sum of "Reg" and "0 − w" will be less than 0x0100, which is the first sum that will set the Carry flag.

It was pointed out to me that the Carry flag in subtraction is really a positive flag: It is set when the result is not less than zero. I find that changing the way I think

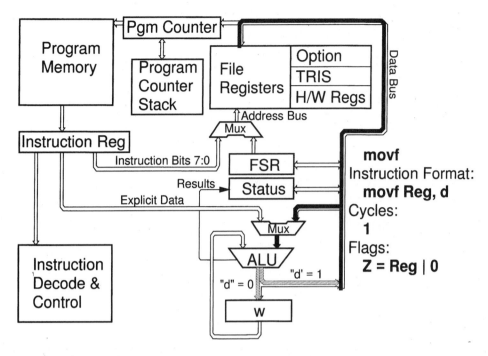

8-4 "movf" instruction.

about Carry when a subtraction operation ("subwf," "sublw," or "addlw 0–") is executed makes predicting how the Carry flag will behave easier.

The Carry flag is also used integrally by the rotate instructions ("rlf" and "rrf"). The instructions rotate the data from the Carry flag through the register and back out to the Carry flag.

The Digit Carry flag works in exactly the same manner as the Carry flag, except it only operates on the lowest four bits (also known as the least significant nybble). Digit Carry is best used in situations where only the least significant four bits of a value are used. The Digit Carry flag is only changed during addition/subtraction instructions.

The program shown in Fig. 8-5 is designed to demonstrate how the flags work during various arithmetic operations.

Throughout the program, you will see that I have put two instruction statements:

```
movlw       0x0F8
andwf       .STATUS
```

into a macro. These two instructions will clear the PIC processor status flags before the next arithmetic operation to better illustrate how the flags are affected before each arithmetic operation. It should be noted that, because the result of the "andwf STATUS" is not zero, the Zero flag will not be set.

8-5 PROG9 Showing how the status flags work.

```
title   "PROG9 - Showing how the status flags work"
;
;  This program plays around with the status flags to show how
;   they are changed by different operations.
;
;  Myke Predko
;  96.05.13
;
  LIST P=16F84, R=DEC
  errorlevel 0,-305
  INCLUDE "\PIC\MPASM\p16F84.inc"


; Macros
ClearFlags Macro                    ; Clear the processor status flags
  movlw     0x0F8
  andwf     STATUS
  endm


; Registers
i           equ 12                  ; Registers to operate on
j           equ 13


  __CONFIG _CP_OFF & _WDT_OFF & _RC_OSC

  PAGE
; Mainline of PROG9

  org       0

  movlw     H'80'                   ; Set "i" to 0x080
  movwf     i
  movlw     8
  movwf     j

; Clear the status flags before doing any operations

  ClearFlags

  movf      i, w                    ; Zero flag not changed

  xorwf     i, w                    ; Zero flag set after instruction

  movf      i, w                    ; Add to show Carry and Zero being
                                    ;   set
  addwf     i, w

  ClearFlags

  movf      j, w                    ; Now, show the Digit Carry flag
  addwf     j, w                    ;   being set

  movf      i, w                    ; Now, Set all three status bits
  subwf     i, w
```

```
        movwf   i

        ClearFlags

        movf    i                   ; Now, just show the Zero flag set

Finished                            ; Finished, just loop around
                                      forever
        goto    $

        end
```

In the program, the first set of instructions is used to initialize the two variables ("i" and "j") that are used in the program. After stepping through these instructions, you will see that the Zero flag is not set.

As you step through this (and any) program, the MPSIM status flags ("F3") will be updated after each instruction. In MPLAB, these three bits are continually displayed with active flags set in uppercase (as is done in the emulator). Unfortunately, MPSIM doesn't have this capability, so you must constantly watch the "F3" and the rightmost three bits (Zero, Digit Carry, and Carry) as you step through a program.

After the variables are initialized, the first operation is to load the "w" register with the contents in the register addressed at "i." As I've noted before, "i" is not an actual variable in the traditional programming sense; it is an address in the file register space. Once "w" is loaded, it is XORed with the same value to produce zero. After this operation, the Zero flag will be set and no other flags will be set.

Next, the contents of the file register space at "i" are added to themselves. Because "i" currently has 0x080, the result will be 0x0100. 0x0100 is greater than an 8-bit number, and the overflow will be signified by the setting of the Carry flag. The least significant 8 bits are put into the result (the "w" register) of the "addwf" instruction. This means that zero is put into the "w" register with the Zero flag set at the same time (along with the Carry flag). The Digit Carry flag is not affected by this operation.

Next, the value at "j" (8) is doubled. The result, 16 (or 0x010 in hex), has an overflow into the next highest nybble. When this happens, the Digit Carry flag is set. The Zero and Carry flags are not set because the result is not equal to zero and it is less than 0x0100.

The next operation, loading "i" and then subtracting it from itself, is going to take some explanation. When the "subwf" operation is executed, we have the following situation:

$$w = 0x080 - w$$
$$= 0x080 - 0x080$$

However, as I've shown before, subtraction is actually addition of a negative. This means that the actual operation is addition of a negative, so the operation becomes:

$$w = 0x080 - ((0x080 \wedge 0x0FF) + 1)$$
$$= 0x080 + (0x07F + 1)$$
$$= 0x080 + 0x07F + 1$$

There are three things happening here that you must be aware of. First, the lower 8 bits of the result are equal to zero (which means the Zero flag is set). The result will be greater than 0x0FF, so the Carry flag will be set. If you look at the lowest nybble, 0x0F plus 1 is 0x010, so the Digit Carry flag will be set. Therefore, all the PIC processor flags will be set after subtracting something from itself.

You might want to execute this operation over several times with different values to prove that this is true.

Once this operation is completed, the zero is stored in "i." Once this is done, the contents of the register at address "i" are run through the ALU to test for zero value. Because "i" is zero, the Zero flag will be set.

With this program, you should get a good idea of how the PIC processor flags work. This is one program which I would suggest that you would be wise to play around with using the different instructions and values and observe how the flags are affected. You might also want to modify it when you develop new applications on your own and are not sure how the flags will behave.

PROG43: Jumping around

In devices where the program store space is larger than the device page, you will have to write code that crosses the page boundary. (See Fig. 8-6.) This program uses the 16C73A instead of the usual 16C84 because of its 4K EPROM program space (the page size is 2K).

As you single-step through the program, you will see execution (and the Program Counter) jump up to the high page (address 0x0800 and above) and then back down again.

By commenting out the "movwf PCLATH" instruction, reassembling, restarting MPSIM, and then single-stepping through, you will see that you will be trapped in an endless loop of the first two instructions ("movlw HIGH HIJump/goto HIJump"). This is because PCLATH has not be changed to point to the page that "HIJump" is located in.

In this example, the incorrect PCLATH value is quite easy to find. Generally, when this happens in an application, it is deep in the execution of the code, and when it jumps off ("branch to the boonies"), it is probably not apparent where the problem occurred or even *when* it occurred. This can mean that you're looking at some serious time with a simulator or emulator.

However, this can be reduced if every interpage "goto"/ "call" is examined first and the PCLATH value is confirmed before starting up the simulator/emulator and stepping through the program.

As noted elsewhere, low-end devices work in a similar manner but use bits in the STATUS register for the page information (there's no explicit PCLATH register, as in the mid-range or high-end PICs).

Data

Data manipulation in the PIC is quite similar to that of other processors. This might have been quite a surprise to you because of the use of the Harvard architecture in the PIC. Actually, the PIC has the same basic addressing modes (immediate, direct, and indexed) that you would find in other processors.

8-6 PROG43 Jumping around.

```
 title  "PROG43 - Jumping around"
;
;  This program jumps around between two different mid-range pages.
;
;  Myke Predko
;  97.02.19
;
 LIST P=16C73A, R=DEC
 errorlevel 0,-305
 INCLUDE ".inc"

__CONFIG _CP_OFF & _WDT_OFF & _RC_OSC

 PAGE
;  Mainline of PROG43

 org            0

 movlw          HIGH HIJump;  Set up PCLATH for "HIJump"
 movwf          PCLATH
 goto           HIJump

loReturn                     ;  Return here from the high page

 goto           $            ;  Finished, loop around forever

 org            0x0800       ;  High page, jump back to low page

HIJump

 clrf           PCLATH       ;  Can clear PCLATH, we know which page
 goto           loReturn

 end
```

The following programs will give you an idea of how the PIC can access data. The programs will also show how the PIC architecture works to actually require *fewer* instructions than other processors to carry out various tasks.

PROG4: Variable manipulation

The most basic form of addressing is by not specifying an address at all. Instead, just the value to use is provided as part of the instruction. This is known as *immediate addressing. Direct addressing* is when the address of the register to be used is actually part of the instruction. Both of these forms of addressing are specified explicitly in the program instructions, and even if the program or the data is changed, these values will remain constant.

Just a note on direct addressed instructions: The address in the instruction is dependent on the bank information in the STATUS register ("RP0" bit). Banks and registers are discussed elsewhere (and are demonstrated in "PROG2" where the same address is used to access the PORTB and TRISB registers).

8-7 PROG4 Variable manipulation.

```
 title  "PROG4 - Variable Manipulation."
;
;  This program is meant to be run in the simulator to show how
;   variables can be manipulated and accessed.
;
;  Myke Predko
;  96.05.08
;
  LIST P=16F84, R=DEC
  errorlevel 0,-305
  INCLUDE "\PIC\MPASM\p16F84.inc"

;  Registers
i    equ 12
j    equ 13

  __CONFIG _CP_OFF & _WDT_OFF & _RC_OSC

  PAGE
;  Mainline of PROG4

  org      0

  movlw    3                   ; Initialized the variables
  movwf    i
  movlw    5
  movwf    j

;  i = i + 1  -  Show how a register is incremented, the result
                 put back

  incf     i                   ; Increment and store the result

;  j = i + 2  -  Now, add two to a register and store it in
   another one

  movlw    2                   ; Get the value to add
  addwf    i, w                ; Add the register to it
  movwf    j                   ; Store the added value

  movf     i, w                ; Now, do the addition a different way
  addlw    2                   ; Load then add, opposite to above
  movwf    j

;  if ( i == j ) then i = i + 1  -  Do a simple comparison with an
                                    add

  movf     i, w                ; Subtract i from j and if the result
  subwf    j, w                ;  is equal to zero, increment i
  btfsc    STATUS, Z
   incf    i

Finished                       ; Finished, just loop around forever
  goto     $

  end
```

PROG4 (Fig. 8-7) carries out a number of immediate and direct instructions. After assembling the code, you can begin to single-step through the program.

The first block of code is the file register initialization. To initialize a register, typically "w" is loaded with the initial value to be put into the file register (using the "movlw" instruction). The contents of "w" are then put into the file register (using the "movlw"/"movwf" instructions). If a register is to be initialized to zero, a "clrf" instruction is usually used. There aren't any simpler ways of initializing the file register. Because the file register is separate from the program space, the initial values of registers or variables cannot be specified as they can be in traditional architectures (e.g., the Intel 8086) that use the same physical memory for both control store and variable (register) store.

Next, incrementing a register is demonstrated. As can be seen, the statement:

```
i = i + 1
```

can be neatly reduced to:

```
incf i, f
```

or (as is used):

```
incf i
```

As noted in chapter 6, the second "incf" instruction is exactly the same as the first.

Many processors have instructions that can carry out a simple increment. What makes the PIC special is when the original high-level statement puts the result into another variable.

For example:

```
j = j + 3
```

In a traditional processor (e.g., the Motorola 68HC11), the following code would be required:

```
ldaa        j        ;  Load the accumulator with the contents of "j"
adda        3        ;  Add three to the value
staa        j        ;  Store the result of the operation
```

In the PIC, the equivalent series of instructions is:

```
movlw       3        ;  Get the value to add
addwf       j        ;  Add the contents of "w" to the value in "j"
```

Along with the PIC code taking fewer instruction addresses than the 'HC11 (2 versus 8 for the three instructions), the PIC also requires only 8 clock cycles versus the 'HC11's 32 (based on 4 "E-Clocks" for every instruction cycle).

The next operation ("j = i + 2") is carried out in two different ways in the PIC code. After single-stepping through each of them, you will see that both get the same result. The first method adds "i" to an immediate value, while the second adds an immediate value to the "w" register (which was loaded with the value at address "i"). Either method is valid, and neither holds advantages over the other.

The last block of code is used to simulate a short block of high-level conditional code. The snippet of code compares two values, and if they are equal (subtracting one from the other results in zero), one of the two values is incremented.

The comparison (subtraction) is carried out with the result being put into "w," so neither value is changed by the operation. The results of the comparison are stored in the STATUS register (Zero, Carry, and Digit Carry flags). The Zero flag is checked for zero (both values being equal). If the Zero flag isn't set, the next instruction is skipped (the variable increment).

Now, if the high-level statement executed on the condition was more than one instruction long, the condition check and skip code would change to:

```
        movf    i, w            ; Get one value for the comparison
        subwf   j, w            ; Compare it to the other, result in "w"
        btfss   STATUS, Z
        goto    IF_Skip         ; Else, skip over the conditional code
        .
        :                       ; Code to be executed if "i" == "j"
IF_Skip
```

PROG5: Simulating a variable array

One of the most useful data constructs is the array. In PIC Assembly language terms, an array can be thought of as *indexed Addressing*. The FSR register is used to access a software specified address (i.e., an address loaded and manipulated in the FSR by software during execution time). Indexed addressing differs from immediate or direct Addressing in that the address is dynamically created during program execution.

PROG5 gives examples of how array reads and writes can be accomplished. (See Fig. 8-8.) The first part of the program is (as always) the variable initialization.

8-8 PROG5 Simulating an array.

```
title   "PROG5 - Simulating an array."
;
;   This program shows how a single-dimensional array can be
;     implemented on a PIC using the FSR and INDF registers.
;
;   Myke Predko
;   96.05.10
;
    LIST P=16F84, F=INHX8M, R=DEC
    errorlevel 0,-305
    INCLUDE "\pic\mpasm\p16F84.inc"

;   Registers
i       equ 12
Array equ 13                    ;   Four bytes of an array

  __CONFIG _CP_OFF & _WDT_OFF & _RC_OSC

    PAGE
;   Mainline of PROG5

    org     0
```

```
        movlw    3                 ;  Initialized the variables
        movwf    i

        movlw    Array             ;  Set up the array pointer to start
        movwf    FSR

        movlw    'm'               ;  Initialize the array
        movwf    Array             ;  Put in "myke"
        movlw    'y'
        movwf    Array + 1
        movlw    'k'
        movwf    Array + 2
        movlw    'e'
        movwf    Array + 3

;  i = Array[ 2 ] - Get the third element in the array

        movlw    2                 ;  Move the array to the character
        addwf    FSR

        movf     INDF, w           ;  Get the character at third
        movwf    i                 ;     element and store it

        movlw    2                 ;  Restore the pointer to start of
        subwf    FSR               ;     the array

;  Array[ 0 ] = 'M' - Change the high byte of the array

        movlw    0                 ;  Move the array to the character
        addwf    FSR

        movlw    'M'               ;  Store the new value
        movwf    INDF

        movlw    0                 ;  Restore the pointer to start of
        subwf    FSR               ;     the array

Finished                          ;  Finished, just loop around forever
        goto     $
        end
```

Once the array is defined in RAM, the general form of accessing an element of the array is:

```
movlw             element#    ;  Get the address of the element to
addwf             FSR         ;     be accessed
read/write Array element      ;  Access the array element
movlw             element#    ;  Restore the array pointer (FSR) back to
subwf             FSR         ;     the start of the array
```

Before and after the array element access, the FSR register is kept at the start of the array so that the program has a constant starting point. Note that, in the second access (writing "M" at the start of the array) has an add and subtract of zero to the FSR. Obviously, this could have been optimized (removal of the "movlw/addwf" and "movlw/subwf" statements), but all the instructions were left in to show that the first element of an array is given the index of zero.

The index into the array can also be another variable (not a constant as shown in PROG5).

For example, if you wanted to create the Assembler code for:

```
k = Array[ i ];
```

the following Assembler would be used (it is assumed that the FSR is already loaded with the address of the start of "Array"):

```
movf     i, w          ; Update the FSR with the index "i"
addwf    FSR
movf     INDF, w       ; Read and save the element Array[ i ]
movwf    k
movf     i, w          ; Restore FSR to point to start of "Array"
subwf    FSR
```

Multidimensional arrays are best created with the various dimensions as a multiple of two. This isn't an absolute requirement, but it will make coding the array easier.

For example, the PIC assembly code for:

```
j = Array[ i ][ 2 ]
```

where "Array" is a 4×4 two-dimensional array would be:

```
bcf      STATUS, C     ; Get i x 4 as the 1st dimension value
rlf      i, w
movwf    ArrayTemp
bcf      STATUS, C
rlf      ArrayTemp
movlw    2             ; Get the 2nd dimensional offset
addwf    ArrayTemp     ;  and add it to the first dimension
movf     ArrayTemp, w  ; Add total offset to the current array
addwf    FSR           ;  pointer
movf     INDF, w       ; Get the byte pointed at Array[ i ][ 2 ]
movwf    j
movf     ArrayTemp, w  ; Restore FSR to its previous value
subwf    FSR
```

There is one thing to watch for when you are using arrays indexed by variables: the array limits. You must make sure that the FSR *never* points outside of the array boundaries.

PROG6: Simulating a stack for arithmetic operations

When a high-level language converts a complex arithmetic expression:

```
i = (j + 2) + (k * 3)
```

it usually uses a stack and executes the instructions in a "push/pop" format, such as:

```
push     j             ; j + 2
push     2
add
push     k             ; k * 3
push     3
mul
add                    ; (j + 2) + (k * 3)
pop      i             ; i = (j + 2) + (k * 3)
```

The actual stack instructions are simply inserted for the stack instructions like macros.

The PIC doesn't have a traditional stack pointer along with "push/pop" operations, but they can be simulated easily as is shown in Fig. 8-9.

8-9 PROG6 Using a stack during arithmetic calculations.

```
title   "PROG6 - Using a stack during arithmetic calculations"
;
;  This program shows how to use a stack for saving intermediate
values during complex mathematical operations.
;
;
;  Myke Predko
;  96.05.10
;
  LIST P=16F84, F=INHX8M, R=DEC
  errorlevel 0,-305
  INCLUDE "\PIC\MPASM\p16F84.inc"

; Registers
a          equ 12              ;  Registers to operate on
bi         equ 13              ;  "b" is an MPASM pseudo-op
c          equ 14
d          equ 15
Stack      equ 16              ;  Stack for operations (two bytes)

  __CONFIG _CP_OFF & _WDT_OFF & _RC_OSC

  PAGE
;  Mainline of PROG6

  org    0

  movlw  3                     ;  Initialized the variables
  movwf  a
  movlw  5
  movwf  bi
  movlw  7
  movwf  c
  movlw  9
  movwf  d

  movlw  Stack                 ;  Set up the stack pointer
  movwf  FSR

;  d = (((a << 1) + bi) << 1) + ((a >> 1) - c)

  rlf    a, w                  ;  Get values in the 1st set of
                                  brackets
  addwf  bi, w
  movwf  INDF
  rlf    INDF
  incf   FSR

  rrf    a, w                  ;  Get the values in 2nd set of
                                  brackets
```

8-9 PROG6 Using a stack during arithmetic calculations. *Continued.*

```
movwf   INDF                    ; Do the temporary storage
movf    c, w                    ; Subtract "c" from the shifted value
subwf   INDF, w

decf    FSR                     ; Finally, add two values together
addwf   INDF, w                 ;   and store
movwf   d

; c = ((Table[ a ] << 1) - (bi - d)) + (37 + d)

movf    a, w                    ; Get the first value
movwf   INDF
call    Table                   ; Get the table element
movwf   INDF
rlf     INDF
incf    FSR

movf    d, w                    ; Now, do the next value
subwf   bi, w                   ; Do the subtraction

decf    FSR                     ; Do the subtraction
subwf   INDF                    ; Store back on the stack
incf    FSR

movlw   37                      ; Get the last value first
addwf   d, w

decf    FSR                     ; Add the two stack values together
addwf   INDF, w
movwf   c                       ; Store the result

Finished                        ; Finished, just loop around forever
  goto  $

; Put in the table subroutine
Table
  movf  INDF, w                 ; Get the parm off the stack
  addwf PCL                     ; Jump to the offset in the stack
  retlw 'a'                     ; Return the specified table element
  retlw 'b'
  retlw 'c'
  retlw 'd'
  retlw 'e'
  retlw 'f'

  end
```

Looking over PROG6, you might feel the code is needlessly complex (and I'd have to agree with you in regard to this application). Where stack-based operations are important is in programs with many complex operations or different data formats (i.e., 8-, 16-, 32-bit or floating point numbers) that need multiple temporary storage registers.

In the very complex mathematical case, using a stack for arithmetic calculations is an advantage because it means much fewer temporary file registers will be required—only what is required will be used. For simpler mathematical requirements, a stack is not the best way to go.

Just one word about compilers. While most compilers must be able to handle very complex operations, they should also be able to optimize statements like:

```
j = j + 1
```

into

```
incf        j
```

rather than process through:

```
push        j
push        1
add
pop         j
```

Subroutines

Using subroutines in the PIC can seem easy, without any major things to worry about. For the most part this is true, but there are some things to watch for. The next three experiments will give you an idea of how subroutines will work and how data is passed between a caller and a subroutine.

PROG7: Passing data through registers

The simplest method of passing data to a subroutine is through a register. In the PIC, this can be done by either passing a byte parameter in the "w" register or using a temporary file register common to both the caller and the routine.

PROG7 (Fig. 8-10) shows how both these methods would be implemented.

8-10 PROG7 Passing subroutine parms via registers.

```
 title  "PROG7 - Passing subroutine parms via registers"
;
;   This program shows how to pass parameters back and forth
between the mainline and routines using registers.
;
;
;   Myke Predko
;   96.05.10
;
   LIST P=16F84, F=INHX8M, R=DEC
   errorlevel 0,-305
   INCLUDE "\PIC\MPASM\p16F84.inc"

; Registers
i          equ 12               ;  Registers to operate on
j          equ 13
Temp       equ 14

   __CONFIG _CP_OFF & _WDT_OFF & _RC_OSC
```

8-10 PROG7 Passing subroutine parms via registers. *Continued.*

```
    PAGE
;   Mainline of PROG7

    org     0

    movlw   3                   ;   Initialized the variables
    movwf   i
    movlw   5
    movwf   j

;   j = increment( i )

;   Pass parameters using the "w" register

    movf    i, w                ;   Get value to increment
    call    increment_w         ;   Increment using the 'w' register
    movwf   j

;   Pass parameters using temporary registers

    movf    i, w                ;   Get the value to increment
    movwf   Temp
    call    increment_Temp      ;   Increment the Temp value
    movf    Temp, w             ;   Store the incremented value
    movwf   j

Finished                        ;   Finished, just loop around forever
    goto    $

;   Increment Subroutines

increment_w                     ;   Increment the value in "w"
    addlw   1                   ;   Add one to value in "w"
    return

increment_Temp                  ;   Increment the Temp value
    movlw   1
    addwf   Temp
    return

    end
```

The first method uses the "w" to store input parameter and return the result. This works well and is quite efficient but does have one major drawback. This is the limitation to 1 byte passed in and out of the routine. For some routines, there isn't any problem; for others, this simply isn't sufficient.

Using file registers to pass data back and forth eliminates the limitation of using "w" to pass a single byte parameter back and forth. However, this method introduces its own problem: When calling multiple nested routines, there is the issue of having correct common parameter registers for the different routines.

PROG8: Passing parameters on a stack

Passing parameters to and from a subroutine on a stack (Fig. 8-11) eliminates the problems noted previously in passing parameters in shared registers.

8-11 PROG8 Passing subroutine parms via stack.

```
 title  "PROG8 - Passing subroutine parms via stack"
;
;  This program shows how to set up a pseudo-stack and use it to
pass parameters between the mainline and routines.
;
;
;  Myke Predko
;  96.05.10
;
   LIST P=16F84, F=INHX8M, R=DEC
   errorlevel 0,-305
   INCLUDE "\PIC\MPASM\p18F84.inc"

;  Registers
i          equ 12           ; Registers to operate on
j          equ 13
Stack      equ 14           ; Start of the virtual stack

   __CONFIG _CP_OFF & _WDT_OFF & _RC_OSC

   PAGE
;  Mainline of PROG8

   org    0

   movlw  4                 ; Initialized the variables
   movwf  i

   movlw  Stack             ; Set up the stack
   movwf  FSR

;  j = Increment( i )

;  Get the value of i + 1 and put it in "j"

   incf   FSR               ; Make space for the sum

   movf   i, w              ; Get first value to add and store on
   movwf  INDF              ;   the stack
   incf   FSR
```

8-11 PROG8 Passing subroutine parms via registers. *Continued.*

```
   call    Increment        ;  Increment the value

   decf    FSR              ;  Get and store the sum
   movf    INDF, w
   movwf   j

Finished                    ;  Finished, just loop around forever
   goto    $

;  Increment subroutine

Increment                   ;  Add the two values on the stack

   decf    FSR              ;  Get the first value and increment it
   incf    INDF, w

   decf    FSR              ;  Store the result
   movwf   INDF

   incf    FSR              ;  Reset the stack pointer

   return

   end
```

In PROG8, you should notice that the FSR always points to the next available stack element and that, for returned parameters, I make space for them *before* loading the input parameters on the stack. This means that the stack is always ready to take a new value without any of the previous values being affected.

When determining the number of file registers to allocate for parameter passing, simply determine the number of parameters (both input and output) for each routine and sum the "deepest" path to get the total stack required. Chances are that you'll find that you'll need fewer stack file registers than parameter registers for each subroutine.

While I feel implementing a PIC stack is more trouble than it's worth in arithmetic operations, they do offer advantages for the user in the area of multiple or nested subroutines.

PROG10: Blowing up the Program Counter stack

The PIC's Harvard architecture means that the PIC uses a separate LIFO (Last In-First Out) register stack for storing the Program Counter when execution changes to subroutines or interrupts. Because the stack is separate from the control store and file registers, it is quite limited.

This limitation means, in the mid-range devices, that no more than 8 levels of subroutines and interrupt handler subroutines can be executed. For most programming styles, this is not a problem. It is an issue if you are going to implement a recursive subroutine (which is a subroutine that can call itself).

What happens when more than 8 subroutines are called? To answer this I have created the program shown in Fig. 8-12.

8-12 PROG10 Bowing up the PC stack.

```
 title   "PROG10 - Blowing up the PC stack."
;
;  This program calls a bunch of routines to show how the stack
;   can be exceeded and an incorrect address returned
;
;  Myke Predko
;  96.05.13
;
 LIST P=16F84, R=DEC
 errorlevel 0,-305
 INCLUDE "..\p16F84.inc"

;  Registers

 __CONFIG _CP_OFF & _WDT_OFF & _RC_OSC

 PAGE
;  Mainline of PROG10

 org    0

 call   Prog1              ;  Now, just call subroutines

Finished                   ;  Finished, just loop around forever
 goto   $

;  Subroutines
;  The subroutines consist of:
;  Prog_at_Label
;    calling Prog_at_label_plus_one
;    return

Prog1

 call   Prog2

 return

Prog2

 call   Prog3

 return

     .                     ;  Repeat the simple subroutines
     :

Prog10

 nop                       ;  At end of "call" chain, start returning

 return

 end
```

After assembling this program and selecting "System Reset," you can begin single-stepping through the program. As you step through the program, you will receive error messages in "Prog8" when "Prog9" is called saying that the stack is overflowing. Keep stepping and you will get another message at the call to Prog10.

Keep stepping through until you reach the "return" statement in Prog10. At this point, the Program Counter stack elements, which had the return address back into Prog1 and Prog2, have been overwritten. As you continue to single-step through the program, you will find, at Prog3's return, you will get a message stating that an underflow has occurred. After clearing this message and single-stepping on, you will see that execution has jumped back to Prog9.

When this happens, it means that the stack has been overwritten and the original contents have been lost. It also means that execution can never return to the original calling point.

To avoid this situation, I always check all the subroutines used and make sure the 8 levels of Program Counter stack are *never* exceeded. This usually means that I find the deepest level in the mainline and add it to the deepest subroutine level plus one more for the interrupt handler (and the deepest nesting level of the interrupt handler's subroutines).

For the most part, the limited Program Counter stack of the PIC is not a great liability. As I've noted, the only programming construct that this limitation makes not recommended for the PIC is recursive subroutines.

Table data

One of the most useful features of the PIC is its ability to read table data. Throughout this book, you will find numerous examples of different ways in which tables can be implemented. However, to use tables effectively, a couple of rules must be followed. PROG14 and the other experiments in this section will show you how to implement tables and avoid any potential problems with them.

PROG14: Table calling and placement

PROG14 (Fig. 8-13) simply loops through a table, storing each value after it's read. Once the table has been read through fully, the PIC goes into an infinite loop.

8-13 PROG14 Table reading.

```
title  "PROG14 - Table reading."
;
;  An invalid table read is implemented.
;
;  Myke Predko
;  96.05.17
;
   LIST P=16F84, R=DEC
   errorlevel 0,-305
   INCLUDE "\PIC\MPASM\p16F84.inc"

;  Registers
```

```
i     equ 12
j     equ 13

__CONFIG _CP_OFF & _WDT_OFF & _RC_OSC

  PAGE
; Mainline of PROG14

  org   0

; Table Read, Get values in Table until they equal zero.

  clrf   i                   ; Use "i" as the index
Table_Loop                   ; Loop around here until finished
  movf   i, w                ; Get the index into the table
  incf   i                   ; Increment the table index
  call   Table               ; Call the table
  movwf  j                   ; Store the Returned value
  movf   j                   ; Check to see if we have zero
  btfss  STATUS,Z            ; Skip if it is set
   goto  Table_Loop          ; If not, then loop around again

  goto   $                   ; All done, go to infinite loop

  org    0x0FC               ; This table goes over a 16-bit boundary
Table
  addwf  PCL                 ; Get the table offset
  retlw  'T'                 ; Return the complete "Table" message?
  retlw  'a'
  retlw  'b'
  retlw  'l'
  retlw  'e'
  retlw  0                   ; End the string

  end
```

Now, when you try to implement and simulate the program, you will find that the table is not fully read and the infinite loop instruction ("goto $") is not executed.

This is because the table goes over the first 256-address boundary (and the "addwf PCL" instruction jumps to address 0, rather than address 0x0100).

This is the biggest potential problem to watch out for when working with tables.

As you step through the program, each table element is read out. You'll see that, when "i" is equal to 3, the table address *should* be 0x0100. However, because the PCLATH register is initialized to zero (and never changed), the address executed after the "addwf PCL" instruction is zero (the "clrf i" instruction).

After executing the program as it stands, comment out the "org 0x0FC" line, re-assemble the code, and step through it again. Now, you will find that the table works properly (because it is totally within the first 256 addresses).

As part of my coding style (which you will see more of as you progress through this chapter and chapter 9), I usually put all my tables at the start of the program. This can usually be done safely, because this means I have well over 240 addresses (if interrupts are used, you have to jump over them as well) with which to put in my tables (usually lots of space).

If the total tables go beyond the first 256 addresses of the PIC program, then some tables must be relocated to another 256 address block (usually the next one) with the following header:

```
HighTable
    movlw       high $          ; Load PCLATH with the current table
    movwf       PCLATH
    addwf       PCL             ; The rest of table is handled normally
    retlw       ...
```

PROG44: Tables longer than 256 entries

Sometimes you will have to create a table that is longer than 256 entries. The table will have to be accessed using sixteen bits (two file registers) for the index into the table rather than the single one used for the small table in the previous section and PCLATH will have to be changed. (See Fig. 8-14.)

8-14 PROG44

```
 title   "PROG44 - Tables longer than 256 entries."
;
;  This program demonstrates how a table that is longer than
;   256 entries (and therefore, jumps over the 256 "addwf PCL"
;   page boarder).
;
;  Myke Predko
;   97.02.19
;
   LIST P=16F84, R=DEC
   errorlevel 0,-305
   INCLUDE "\PIC\MPASM\p16F84.inc"

 __CONFIG _CP_OFF & _WDT_OFF & _RC_OSC

;  Registers
 CBLOCK 0x00C
Tbllo, TblHI                     ; Offsets into the table
Temp
 ENDC

   PAGE
;  Mainline of PROG44

   org     0

   goto    MainLine              ; Skip Over the table


MSGTable                         ; Place all the messages here
   movlw   HIGH TableStart       ; Set up PCLATH for the table
   addwf   TblHI, w
   movwf   PCLATH
```

```
       movlw   TableStart & 0x0FF    ; Figure out the offset
       addwf   Tbllo, w
       btfsc   STATUS, C             ; If necessary, increment PCLATH
        incf   PCLATH                ;  to get the correct 256-address
                                     ;  page
       movwf   PCL                   ; Update the PC

TableStart                          ; Table data
Msg0
       dt      "This is the First Message", 0
Msg1
       dt      "This is the Second Message", 0
Msg2
       dt      "This is the Third Message", 0
Msg3
       dt      "This is the Fourth Message", 0
Msg4
       dt      "This is the Fifth Message", 0
Msg5
       dt      "This is the Sixth Message", 0
Msg6
       dt      "This is the Seventh Message", 0
Msg7
       dt      "This is the Eigth Message", 0
Msg8
       dt      "This is the Nineth Message", 0
Msg9
       dt      "This is the Tenth Message", 0


MainLine                            ; Execute the program

       movlw   9                    ; Get the 10th message
       call    GetMSG

       goto    $                    ; "Tbllo/TblHI" have the offset
                                    ;  to the 10th message


GetMSG                              ; Figure out where the specified
                                    ;  message is
       movwf   Temp                 ; Save the message count

       clrf    Tbllo                ; Reset the table values
       clrf    TblHI

GM_Loop

       movf    Temp                 ; If Temp == 0, then stop search
       btfsc   STATUS, Z
        goto   GM_End

       call    MSGTable             ; Get the value in the table

       iorlw   0                    ; Are we at the end of a message?
       btfsc   STATUS, Z
        decf   Temp                 ; Yes, decrement Temp
```

8-14 PROG44 *Continued.*

```
incf   Tbllo                    ;  Point to the next table element
btfsc  STATUS, Z
 incf  TblHI

goto   GM_Loop

GM_End                          ;  Tbllo/TblHI now point to the
                                ;    first character in the
                                ;    specified message

return

end
```

When "GetMSG" returns, "Tbllo/TblHI" will be pointing to the first entry in the appropriate message.

The big thing to note about this program over the previous one is that a new PCL and PCLATH are calculated before the "MSGTable" entry can be retrieved. In the previous program, the offset of the desired table element is simply added to the off-set of the current PCL. As can be seen, this is not a very hard program to implement, but there are some rules to follow to make sure both PCL and PCLATH have the correct values in the table.

There are also a couple of minor things to note in PROG44. First is the use of the "CBLOCK" pseudo-op, which makes allocating register values easier. Rather than having to make equates for each register (and moving others, if needed), the "CBLOCK" statement will equate each label to an incrementing value.

The second, and more profound, change with respect to the previous program is the use of the "dt" pseudo-op, which puts in a "retlw Const" instruction for each entry on the line. This greatly simplifies (and shortens) the program and makes the source much easier to read.

As you go through the programmer/emulator code (as well as some of the projects), you will see that the code presented here is the basis for providing a simple interface to a list of messages. This method of doing it is quite efficient in terms of providing a selection of messages that can be indexed to in the least amount of space.

PROG12: Execution control state machines

Execution state machines are a table application that can offer some advantages in simplifying your programs in certain circumstances.

If you were keying off a file register value (e.g., 0, 1, 2, etc.) you could use the following code:

```
movf   Reg, w
btfsc  STATUS, Z
 goto  State0                   ;  Reg = 0
addlw  -1
btfsc  STATUS, Z
 goto  State1                   ;  Reg = 1
```

```
addlw           -1
btfsc           STATUS, Z
  goto          State2                      ;  Reg = 2
  .
  :                                         ;  ... And so on ...
```

However, a table would make the branching on the contents of "Reg" much simpler:

```
movf            Reg, w                      ;  Get the Index
addwf           PCL
goto            State0                      ;  Reg = 0
goto            State1                      ;  Reg = 1
goto            State2                      ;  Reg = 2
  .
  :                                         ;  ... And so on ...
```

There are three advantages to using a table in this situation:
- Fewer instructions are required (for the initial case, three instructions are required for each case; in the table case, one instruction is required for each case, plus two overhead bytes).
- Each different jump condition takes the same number of cycles. This could be important for some applications.
- The table jumps are a lot easier to understand by looking at.

Because of these features, table jumps are good for state machines.

I define a state machine as a program that uses a single value for executing different functions the program. This value is set according to the previous state and current environmental conditions (e.g., "if RB0 is set, increment the state and execute the response to RB0 being set"). In many ways, a state machine is an exercise in *nonlinear* programming.

I realize that PROG12 (Fig. 8-15) is a pretty simple example of a state machine, but it does show how the state is changed with different conditions and the program progresses forward.

8-15 PROG12 Demonstrating a state machine.

```
 title  "PROG12 - Demonstrating a state machine"
;
;  This program demonstrates how a state machine could work
;   with the PIC architecture.
;
;  Myke Predko
;  96.05.14
;
  LIST P=16C84, F=INHX8M, R=DEC
  errorlevel 0,-305
  INCLUDE "\PIC\MPASM\p16F84.inc"

;  Registers
i             equ 12          ;  General counter
state         equ 13          ;  Returned value
Temp          equ 14          ;  Temporary storage variable
```

8-15 PROG12 Demonstrating a state machine. *Continued.*

```
__CONFIG _CP_OFF & _WDT_OFF & _RC_OSC

    PAGE
;   Mainline of PROG12

    org     0

    clrf    i                   ; Initialize variables
    clrf    state

    clrf    PORTB               ; Set up PORTB
    bsf     STATUS, RP0
    clrf    TRISB & 0x07F
    bcf     STATUS, RP0

;   Now, execute the program

Loop                            ; Return here after every execution
    movlw   1                   ; Check least significant bit of PORTB
    andwf   PORTB, w
    movwf   Temp
    bcf     STATUS, C           ; Now, shift over the state variable
    rlf     state, w
    addwf   Temp, w             ; Add least significant bit of PORTB

    addwf   PCL                 ; Jump to the correct state execution
                                ;   vector
    goto    State0
    goto    State0
    goto    State10
    goto    State11
    goto    State2
    goto    State2

;   State Routines...

State0                          ; Increment i to 4
    incf    i, w
    movwf   i

    sublw   3                   ; Is "i" greater than 3?
    btfss   STATUS, C
    incf    state               ; Yes, increment the state variable

    goto    Loop                ; Execute the state value again

State10                         ; Increment the LSB of PORTB if == 0
    movlw   1
    addwf   PORTB

    goto    Loop

State11                         ; Shift PORTB by one until Carry set
```

```
    bcf     STATUS, C
    rlf     PORTB

    btfsc   STATUS, C       ;  Is the Carry set?
      incf  state           ;  Yes, go to the next state

    goto    Loop

State2                      ;  Reset everything and restart program
    clrf    i
    clrf    state

    goto    Loop

    end
```

State machines are particularly useful in low-end PICs, where the two-level stack might be a hindrance to traditional programming methods.

Playing with the hardware

The programs given in the rest of this chapter are designed to run on actual hardware. Before we begin working with the hardware there are two things that I want to mention.

The first is that now is the time the programmer should be attached to your PC. As I've noted elsewhere in the text, there are a plethora of different programmer designs out there for the 16C84/16F84. From here on, I will reference the programmer presented in this book. This device interfaces directly to a wide variety of host systems, and after the program is "burned" into the PIC, the program can be executed directly. This is by far the easiest method of developing PIC applications along with changing/updating programs into PICs. (Figure 8-1 showed the programmer plugged into a protoboard.)

The second is even more central to the creation of the experiments presented in this chapter.

If you aren't using my programmer design, you can use the same type of protoboard, as described at the start of this chapter. I also use a very basic central core to each circuit, which is used in all the remaining programs. This central core consists of the PIC (I put it in a ZIF socket to prevent damaging protoboard or PIC pins with all the plugging/unplugging that will be done). Power is provided from the protoboard rails and is decoupled, the _MCLR (Reset) pin is attached to Vdd through a 4.7K resistor (more about this later), and a 1-MHz crystal is used for the clock along with two 33 pF capacitors. The circuit looks like Fig. 8-16.

The clocking, power, and reset functions are all provided on the programmer circuit provided in this book.

The reason why I put in the 4.7K resistor pull-up on _MCLR is to allow simple resetting of the PIC. By placing a momentarily "On" switch (labelled "RST" in Fig. 8-16) between _MCLR and ground, you can reset the program without having to power down. This is a very useful feature when you are debugging applications. Because

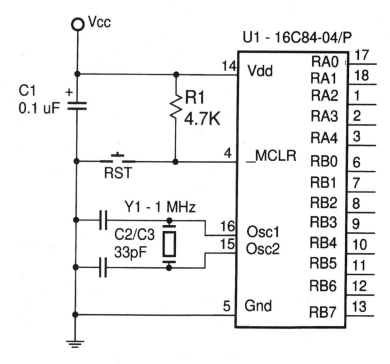

8-16 Basic experiment core circuit.

the _MCLR pin is attached to Vdd, I also specify the power-up timer in the configuration fuses ("_PWRTE_ON" in the "__CONFIG" statement).

Once this core is working, it can be used for all the following experiments.

In some of the programs, this basic circuit will be modified, but this is an excellent place to start and will reduce the hardware creation time significantly.

Note that all the circuit schematics in this chapter reference the "16C84" and not the "16F84". The circuits and code will work with both devices (although, as discussed earlier in this book, the 16F84 should be used).

PROG2: Programming the PIC/turning on a light-emitting diode (LED)

For the first program to be run on actual hardware, I tried to make it as simple as possible because you will probably spend an inordinate amount of time getting your programmer running and the program burned into the PIC. The hardware required is really just a LED attached to one of the I/O ports of the PIC (which is actually the PIC core described earlier). (See Fig. 8-17.)

The configuration of the LED is pretty standard. There are two things, however, that you should notice. The first is that a current-limiting resistor is used. If you read the PIC 16F84 datasheet, you will see that it can only source/sink about 20 mA. While a current-limiting resistor isn't absolutely necessary, putting one in is good design practice.

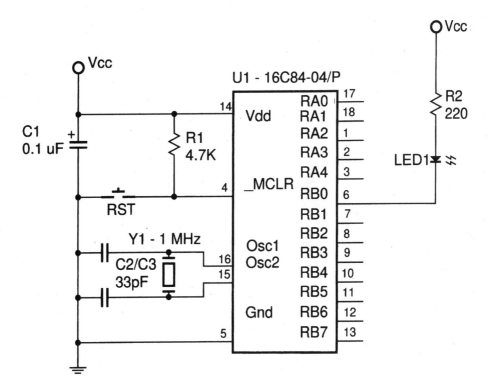

8-17 PROG2 schematic.

The second thing to notice about how the LED is wired up is that the PIC sinks (pulls the pin to ground) rather than sources (drives the pin to Vcc) the current. This is because the PIC can sink more current than it can source. I make as much of my designs negatively active as possible to take advantage of this aspect of the PIC.

The code itself to drive the LED was reviewed earlier in this chapter to show how registers in Bank 1 are accessed. (See Fig. 8-3.)

All the code does is set the PORTB LED bit (RB0) to be output low and then enable the bit for output. Once this is complete, an infinite loop is executed, while the LED is turned on.

Once you've assembled PROG2, you are ready to program ("burn the code into") the PIC. If my programmer is used, after the PIC is in the socket, select "p" (after first making sure the prompt has an "f" to indicate that a Flash/EEPROM part is to be programmed) and download the program into the programmer.

There are a few hints I can offer to speed up your debugging the operation of a simple (cheap) programmer.

When your programmer seems to be working and PROG2 is selected for burning, put in a PIC into the programmer's socket that you haven't yet tried to program and "Verify" its contents. Regardless of what was on the PIC, it should miscompare with PROG2. If it does not, then there is a problem with the programmer, its software, or both. This is usually called an *acid test*, because it makes sure the hardware is working before you use it.

Another acid test you can try is to run the "Verify" function without a PIC in the socket. This should fail as well.

Once the acid test passes (by failing the contents of the PIC), you can burn in the program. Make sure you verify that the PIC has the correct program after the code has been loaded. If you *really* want to make sure the programmer is working correctly and the program has been loaded into the PIC, you can end the programmer's software and power down and disconnect the programmer. After reconnecting the programmer and booting its software, the PIC will pass another "Verify" (acid) test.

Now that you've burned a good PIC and built the circuit, you're ready to install the PIC into the circuit and apply power. Once you do apply power, all that will happen is the LED will light up.

If you put a logic probe on all the PIC I/O pins, you should find that only RB0 (pin 6) should activate the logic probe. Actually, you might want to prowl around a bit with a voltmeter and/or logic probe and see what's happening. You should check _MCLR, Vdd, and Vss and see if the oscillator is running (you'll need a logic probe or

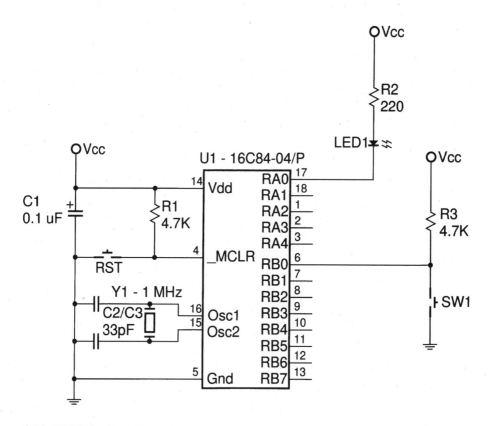

8-18 PROG16 schematic.

oscilloscope to see the oscillator running). If you put a logic probe on "OSC1," you might find that the oscillator stops running. This is due to the additional capacitance of the probe. This means that you should only probe OSC2 to see if the oscillator is running.

While, in the next experiments in this chapter, the purpose is to show how the PIC works in different situations, the purpose of this experiment was to get the programmer and the PIC core (used in all the subsequent experiments) up and running.

PROG16: Polling an input bit

To create a simple input/output system, a momentarily "On" switch is added to a slightly modified PROG2 circuit (Fig. 8-18).

PROG16 (Fig. 8-19) will poll the input bit (RB0) and pass the value to output bit (RA0). When the input line is low, RA0 will be low, allowing the LED to light.

8-19 PROG16 Poll a bit.

```
title   "PROG16 - Poll a bit"
;
;  This is program reads PORTB.0 and stores the result in PORTA.0.
;
;  This program is an enhancement to PROG2.ASM
;
;  Hardware notes:
;    Reset is tied directly to Vcc, and PWRT is enabled.
;    A 220-ohm resistor and LED are attached to PORTA.0 and Vcc
;    A 4.7K pull-up and a switch pull-down are attached to PORTB.0
;
;  Myke Predko
;  96.05.20
;
   LIST P=16F84, F=INHX8M, R=DEC
   errorlevel 0,-305
   INCLUDE "\PIC\MPASM\p16F84.inc"

;  Registers

   __CONFIG _CP_OFF & _WDT_OFF & _XT_OSC & _PWRTE_ON

   PAGE
;  Mainline of PROG16

   org    0

   bsf    STATUS, RP0
   bcf    TRISA & 0x07F, 0   ;  Set RA0 to output
   bcf    STATUS, RP0
```

8-19 PROG16 Poll a bit. *Continued.*

```
Loop                           ;  Loop around looking between the bits
  movlw   1                    ;  Look at the LSB of PORTB
  andwf   PORTB, w
  movwf   PORTA               ;  Put it into the LED on PORTA
  goto    Loop

  end
```

Before modifying the circuit, you might want to try executing PROG16 in MPSIM with the following stimulus file:

```
!  PROG16.STI - Stimulus file for showing how PROG16 works
!
!  This file inputs a low value to the input at Port "B"
!
!  Myke Predko
!
!  96.05.20
!
STEP    RB0
1       1                 ! Put the value high
20      0                 ! Lower the value when inside the infinite loop
40      1                 ! Raise the input line again
```

If you look at PROG13's "MPSIM.INI," you'll see that the stimulus load is done when the program is loaded. As you single-step through the program, you will also see RB0 change from high to low when the count hits 20 instructions. When this happens, the bit written into PORTA (and RA0) is low, and in the actual hardware, the LED turns on.

One of the big advantages of MPSIM is its ability to have simulator instructions (i.e., set registers to specific values, set the program counter, set breakpoints, etc.) embedded in the "MPSIM.INI" file. This can allow you to develop very specific test cases for your code and allow you to avoid having to step (or run) through sections of code that probably will take literally hours to get through.

PROG13: Changing an output bit value inadvertently

Throughout this book, I've talked about being careful about inadvertently changing output port bits.

PROG13 (Fig. 8-20), which uses the same circuit as PROG16, shows how this can happen.

8-20 PROG13 Change a bit inexplicitly.

```
 title  "PROG13 - Change a bit inexplicitly."
;
;  This is a program that will change the output of a bit
;    inadvertently (something discussed in the text).  PORTB.7 is
;    used as the bit to check.
```

```
;
;   This program is an enhancement to PROG16.ASM
;
;   Hardware notes:
;    Reset is tied directly to Vcc, and PWRT is enabled.
;    A 4.7K pull-up and a switch pull-down are attached to PORTB.0
;    A 4.7K pull-up is attached to PORTA.0
;
;   Myke Predko
;   96.05.20
;
    LIST P=16F84, F=INHX8M, R=DEC
    errorlevel 0,-305
    errorlevel 1,-224
    INCLUDE "\PIC\MPASM\p16F84.inc"

;   Registers

    __CONFIG _CP_OFF & _WDT_OFF & _XT_OSC & _PWRTE_ON

    PAGE
;   Mainline of PROG13

    org     0

    movlw   0x01E           ; Initialize PORTA TRISA value in "W"

    clrf    PORTA           ; Clear all the bits in PORTAA

    tris    PORTA           ; Set RA0 as output

Loop                        ; Loop around looking between the bits
    btfsc   PORTB, 0        ; Wait for port B's switch pressed
    goto    Loop

    iorlw   1               ; RA0 is now an input bit
    tris    PORTA

    bsf     PORTA, 2        ; Modify an unrelated bit

    andlw   0x01E           ; Return RA0 to output
    tris    PORTA

    goto    $               ; Finished, loop forever

    end
```

When this program is run, the LED will initially be lit (RA0 is initialized to low); however, after pressing the button, the LED will go out because RA0 has changed to a high logic level, even though it has never been explicitly written to!

To explain what has happened, I will go through the code. You should note that I have used "tris" instructions instead of changing the bank to update the "TRISA" register.

The first three instructions:

```
movlw        0x01E
clrf         PORTA
tris         PORTA
```

set up "w" with the current "TRISA" value and reset all the output bits in PORTA before enabling RA0 as an output.

After this, the PIC waits for a button press (RB0 going low), which causes the program to make RA0 an input bit and then modify another bit in PORTA. When RA0 is changed to an input bit, the LED and resistor will put a high value on RA0 (because there is no current flow).

The next instruction ("bsf PORTA, 2") is what changes the state of the RA0 output bit in the PORTA register.

The "bsf PORTA, 2" instruction can be written as:

```
PORTA = PORTA | (1 << 2)
```

which means that, before bit 2 can be set, PORTA has to be read. In doing this, the hardware pin states are actually read, not the PORTA output register. In this case, the RA0 output latch is at a "0" state, but the pin is at a "1" because of the high voltage caused by the LED/resistor combination.

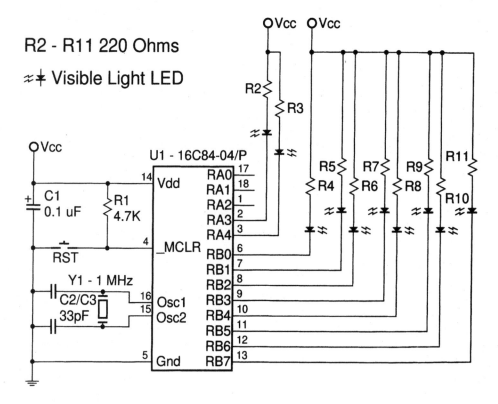

8-21 PROG17 schematic.

This "1" is read from the pin and then programmed back in at the end of the "bsf" instruction. Now, the output latch is set high, which is what is driven when the bit is re-enabled as an output.

Looking through this code, you're probably thinking that this is probably a pretty artificial example and very unlikely during regular operation of the PIC. Actually, this can happen in quite a few different cases, such as I²C communications, where two devices can drive the same line and the PIC has to change a bit type from output to input and back again.

It is actually quite easy to avoid any problems in this situation (just reload the I/O port with the correct values after reading bits and before the outputs are re-enabled).

PROG17: Power/decoupling problems

This next circuit (Fig. 8-21) will be used for the next few experiments to both expand on the level of execution the PIC is running at as well as show how some simple things can make things go awry.

PROG17 (Fig. 8-22) is an eye-pleasing LED display where the PORTA LED's present the _TO and _PD STATUS bits and PORTB will flash 8 LEDs with a progressing pattern.

8-22 PROG17 Decoupling problems.

```
 title  "PROG17 - Decoupling problems."
;
;  This program runs through a bunch of LEDs to show what
;  happens when there is a Vcc programming voltage problem.
;
;  This program is an enhancement to PROG16.ASM
;
;  Hardware notes:
;   Reset is tied directly to Vcc, and PWRT is enabled.
;   A 220-ohm resistor and LED are attached to all the PORTB bits
;   A 220-ohm resistor and LED are attached to RA3 and RA4 to show
;   what
;   _TO and _PD are upon execution start after reset
;
;  Myke Predko
;  96.05.20
;
   LIST P=16F84, F=INHX8M, R=DEC
   ERRORLEVEL 0,-305
   INCLUDE "\PIC\MPASM\p16F84.inc"

; Registers
Temp      equ 12                   ;  16-bit dlay variable
Port      equ 14                   ;  Value saved for the port

   __CONFIG _CP_OFF & _WDT_OFF & _XT_OSC & _PWRTE_ON
```

8-22 PROG17 Decoupling problems. *Continued.*

```
     PAGE
;  Mainline of PROG17

     org    0

     comf   STATUS, w           ;  Display _TO & _PD
     movwf  PORTA

     movlw  0x0FF
     movwf  PORTB               ;  Turn off all the display LEDs
     movwf  Port

     bsf    STATUS, RP0
     clrf   TRISB & 0x07F       ;  Set all the PORTB bits to output
     movlw  0x007               ;  Use PORTA to show _TO and _PD
     movwf  TRISA & 0x07F
     bcf    STATUS, RP0

Loop
     call   Dlay                ;  Now, turn off the LEDs
     movlw  0x0FF               ;  Set them all high
     movwf  PORTB
     call   Dlay                ;  Delay before changing values
     btfsc  Port, 7             ;  Check to see if the Carry is set
      goto  Loop_S
     bsf    STATUS, C
     goto   Rotate
Loop_S
     bcf    STATUS, C
Rotate
     rlf    Port                ;  Now, shift the data over (change
                                ;   Carry)
     movf   Port, w             ;  Save the new value
     movwf  PORTB
     goto   Loop


;  Dlay routine - Delay a half second before returning

Dlay

     ifndef Debug               ;  If Debug NOT defined
     movlw  123                 ;  Actual set up of the delay value
     movwf  Temp
     movlw  30
     movwf  Temp + 1
     else                       ;  If Debug defined
     movlw  2                   ;  Programming Debug delay value
     movwf  Temp
     movlw  1
     movwf  Temp + 1
     endif

D_Loop                         ;  Loop around here until complete
```

```
decf    Temp                    ;  Decrement the high value
btfsc   STATUS, Z               ;  Is the Zero flag Set?
 decf   Temp + 1
movf    Temp, w                 ;  Are we at zero for both?
iorwf   Temp + 1, w
btfss   STATUS, Z
 gotoD_Loop

return

end
```

You'll notice in the code that I have put in the "Debug" pseudo-label. If this label is defined at assembly time, the delay timer will be drastically cut down (only looping 258 times as opposed to 7803 times for normal operation). When debugging with MPSIM (or MPLAB), this will be a very useful option. In the normal operation, there is a half second between changes to the LEDs. This would mean that it would take literally hours for each half-second delay routine to execute.

For this case, even if you are using the programmer presented in this book, you should build the circuit in Fig. 8-21 on a protoboard. The reasons why will become obvious very soon.

Once the circuit is working and you have an eye-pleasing display, pull C1 (the PIC's decoupling cap) from the board.

You should find that the PIC runs for a few iterations and then starts over (I found no more than 5 LEDs would light with C1 pulled out).

When the LEDs go on, there is a current surge (called a *transient*) going through the PIC. The more LEDs that are turned on/off, the greater the transient. At some point (for me, it was 5 LEDs), the transient inside the PIC was so great that the PIC was reset.

When you are running this, note that the _TO/_PD LEDs both stay on, which indicates that a _MCLR or power-on reset (as opposed to a WDT reset) has taken place.

If you put C1 back in while the program is running, you'll see that the PIC now runs properly.

So what did we learn here?

If your answer was that you learned that you can't drive more than 5 LEDs with a PIC, go to the back of the class.

The right answer is that the PIC must have proper decoupling to ensure that it runs without any problems or unexpected resets.

PROG45 (later in this chapter) switches the LEDs in exactly the same manner as PROG17, but the delay loop is carried out by TMR0, and upon overflow, an interrupt handler does the LED switching. PROG45 is an example of how the PIC can do two things at the same time by the use of interrupts.

PROG17: Different oscillators

PROG17, because of the time it takes, can be used to time different oscillators. Actually, the only oscillators I wanted to check on were RC-produced frequencies.

The "nominal" case is obviously running with a 1-MHz clock (and a 4.0-μsec instruction cycle).

In this book, I talk a lot about tolerances, but I wanted to see what I would actually get so I changed PROG17 to use an RC clock (changed the "__CONFIG" parameter to "_RC_OSC" from "_XT_OSC") and put on a 4.7K resistor and 100-pF cap. In one of the Microchip RC oscillator frequency charts, it shows that the PIC will run at just over 1-MHz with a 5K resistor and 100-pF cap. Figure 8-23 shows the RC oscillator frequencies for a 100pF cap for the 16C61 PIC.

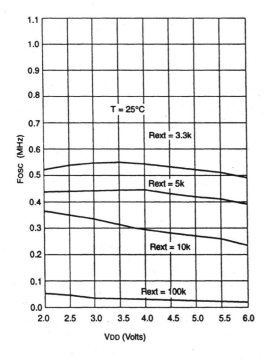

8-23
Typical RC oscillator frequency versus Vdd (Cext = 100 pF).

When the PIC was run and I looked at the divide-by-four clock (instruction clock) coming from OSC2, I found a 4.10-μsec clock period (which translates to a 975-KHz PIC clock frequency). I had expected about a 1.1-MHz to 1.2-MHz clock frequency. In this example, I had a difference of about 15% (which is within the tolerance of 20% that I expected).

This is probably a very good example of what can be expected from an RC clock. I don't have equipment for measuring the precise values of the 4.7K resistor and 100-pF capacitor that I used, and I feel that it would have be a waste of time to precisely set the values for an exact clock frequency. This isn't the point of using an RC oscillator.

An RC clock should only be used in applications where absolute timing doesn't matter.

PROG17: The watchdog timer

One of the most insidious problems you can encounter and have to try to debug is what happens when the watchdog timer is inadvertently enabled.

In the "__CONFIG" line of PROG17, change the "_WDT_OFF" parameter to "_WTD_ON," recompile, and burn the program into a PIC.

Now, when PROG17 is run, it will run for a few seconds and then reset (but with the _TO LED off) and start executing again. This is because the OPTION register in its default condition (0x0FF) has the prescaler set to a full wait, and it is used by the watchdog timer. If the prescaler was directed toward TMR0, then the minimal 18-msec watchdog timer delay would be active, which would make the PIC appear like it was completely dormant.

Now, you have two reasons why the PIC might appear to stop suddenly and reset itself. When this happens, you should make sure that the WDT isn't on (although, if it's supposed to be, you should check for the "clrwdt" instruction) or check to see that there is proper decoupling on the PIC.

PROG21: Reset

Sometimes in an application, resetting the PIC and shutting it down will only be temporary (to allow other hardware to work or to reduce the power required by the application). In these types of applications, the PIC could have information that is needed when it "wakes" up from reset.

As was seen in the previous section, a WDT reset sets the _TO and _PD bits in a certain state, but these two bits don't give enough information to determine whether or not the reset was caused by a power-up or a _MCLR reset.

To help differentiate between the different cases, a variable is typically set up with a pattern that can be checked upon the next reset. If these values match the expected, then the PIC has operated before and certain other values should not be written over. (See Fig. 8-24.) This program can use the same circuit as PROG17. In newer PICs, the "PCON" register will allow differentiating whether reset was caused by –MCRL or Powerup.

8-24 PROG21 Playing around with reset.

```
  title "PROG21 - Playing around with reset."
;
; This program keeps track of whether or not reset vector
;   execution is due to a reset or a power-up.
;
; LEDs with 220-ohm pull-ups on PORTB,1:0
; Switch pulling down MCLR with a 4.7K resistor in series.
;
; Myke Predko
;   96.06.03
;
  LIST P=16F84, F=INHX8M, R=DEC
  errorlevel 0,-305
  errorlevel 1,-224
  include "\PIC\MPASM\p16c84.inc"
  __CONFIG _CP_OFF & _PWRTE_ON & _XT_OSC & _WDT_OFF
; Registers
MASK1     EQU 0x0C                ; Registers for indicating whether
MASK2     EQU 0x0D                ;   or reset or power-up
```

8-24 PROG21 Playing around with reset. *Continued.*

```
MASK3      EQU 0x0E
MASK4      EQU 0x0F
 PAGE
; Code start
   org     0                      ; Return here on power-up

   clrf    PORTB

   movf    MASK1, w               ; Result should be equal to 0x01
   xorlw   0x0AA
   btfss   STATUS, Z
    goto   Power_Up               ; Doesn't, jump to Power_Up rtn
   movf    MASK2, w
   xorlw   0x055
   btfss   STATUS, Z
    goto   Power_Up
   movf    MASK3, w
   xorlw   0x0F0
   btfss   STATUS, Z
    goto   Power_Up
   movf    MASK4, w
   xorlw   -0x00F
   btfsc   STATUS, Z
    goto   Reset                  ; Yes, go to the reset up code

Power_Up                          ; Set the values and indicate with
                                  ;   the output bit

   movlw   0x0AA                  ; Put in the patterns into "MASKn"
   movwf   MASK1
   movlw   0x055
   movwf   MASK2
   movlw   0x0F0
   movwf   MASK3
   movlw   0x00F
   movwf   MASK4

   movlw   0x0FE                  ; Indicate with RB0 first time
   tris    PORTB                  ;   running

   goto    $                      ; Now, just loop around forever
; Reset, indicate with its output bit

Reset

   movlw   0x0FD                  ; Indicate with RB1 that reset
   tris    PORTB                  ;   happened

   goto    $

   end
```

Once it is burned into a PIC and the PIC is run, the LED at RB0 will be lit. This will indicate that the "MASK" values weren't the expected values for a reset restart. Bringing "Reset" down to ground level and then releasing it will be a "_MCLR cycle," and the LED at RB1 should be lit afterward. In the second restart, the "MASK" values were the expected values for a _MCLR reset.

Looking over the code, you might be tempted to change the "MASK" compare to:

```
movf        MASK1, w
xorwf       MASK2, w
xorwf       MASK3, w
xorwf       MASK4, w
btfsc       STATUS, Z
 goto       Reset
Power_Up
```

because all the values XORed together equal zero.

While this might work in some circumstances, it will not be as effective a test as checking the value in each file register individually. This is because there are a lot of bit patterns that equal to zero when XORed together. As will be seen in the next section, some registers have a predisposition to come up in specific states. Both these factors make the simpler code much less reliable in guaranteeing that the PIC is being reset as opposed to being simply reset.

PROG18: Register power-up values/button press with debounce

Throughout this book, I have talked about how register contents are unknown at power-up. I thought it would be interesting to create a small experiment to see what the actual power up values are.

To do this, I slightly modified the PROG17 schematic to the one shown in Fig. 8-25. This new circuit will be used for the next few experiments (which will explain why the switch is at RB0 and the low bit of the output value is at RA0).

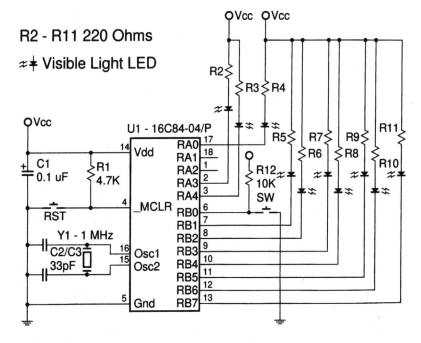

8-25 PROG18 schematic.

PROG18 (Fig. 8-26) will go through each file register of a 16C84 and display them on LEDs attached to RA0 and RB1 through RB7 and wait for a button press to display the next file register's contents. After all 36 registers have been displayed, the PIC goes into an endless loop.

8-26 PROG18 Register contents.

```
 title  "PROG18 - Register contents."
;
;  This program reads the value in a RAM register and outputs it
;  inverted onto PORTB (which has LEDs to display the value). All the
;  RAM registers are read and displayed. A button is used as the
;  instigator of the next value read. The FSR is copied into the LEDs
;  to display the current register being displayed.
;
;  This program is a modification of PROG17.ASM
;
;  Hardware notes:
;   Reset is tied directly to Vcc, and PWRT is enabled.
;   A 220-ohm resistor and LED are attached to PORTA.1:0
;   A 4.7K pull-up and switch pull-down are attached to PORTA.4
;   A 220-ohm resistor and LED are attached to all the PORTB.7:0
;
;  Myke Predko
;  96.05.20
;
 LIST P=16C84, F=INHX8M, R=DEC, N=45
 errorlevel 0,-305
 errorlevel 1,-224
 INCLUDE "\PIC\MPASM\p16c84.inc"

;  Registers
Reg       EQU 0x0C                    ;  Start of register area


 __CONFIG _CP_OFF & _WDT_OFF & _XT_OSC & _PWRTE_ON

 PAGE
;  Mainline of PROG18

 org    0

 movlw  0x0FF
 movwf  PORTB                    ;  Turn off all the indicator LEDs
 movwf  PORTA
 bsf    STATUS, RP0
 movlw  1                        ;  Set up the output bits (everything
 movwf  TRISB & 0x07F            ;   but RB0)
 movlw  0x01E
 movwf  TRISA & 0x07F            ;  Just RA0 is an output on PORTA
 movlw  0x0D4                    ;  Set up TMR0 for use in ButtonPress
 movwf  OPTION_REG & 0x07F
 bcf                STATUS, RP0

 movlw  Reg                      ;  Set up index to the start of the regs
 movwf  FSR
```

```
Loop                            ; Loop to here for each register

    comf   INDF, w              ; Get the bits in "Reg" and display
    movwf  PORTB                ;  so the high bit LEDs are on
    movwf  PORTA                ; Just load RA0 with the "Reg" value

    incf   FSR                  ; Point to the next register read

    movf   FSR, w               ; Are we at end of the register space?
    addlw  0 - 0x030

    btfsc  STATUS, C            ; If Carry is NOT set, then end
    goto   Finished

    call   ButtonPress          ; Wait for button press and released

    goto   Loop

Finished                        ; Program is complete, loop forever
    goto   $

; ButtonPress routine - Wait for the button to be pressed and released
ButtonPress

BP1                             ; Wait for the button to be pressed
  btfsc  PORTB, 0
    goto   BP1

    clrf   TMR0                 ; Now, will 30 msec go by in this state?
BP1_Loop                        ; Loop here until 30 msec go by (TMR0 ==
    comf   TMR0, w              ;   0x0FF) or button "bounce"
    btfsc  TATUS, Z
    goto   BP2                  ; TimeOut, wait for release
    btfss  PORTB, 0
    goto   BP1_loop

    goto   BP1                  ; Button high, it bounced, wait again

BP2                             ; Now, repeat above for button release
  btfss  PORTB, 0
    goto   BP2

    clrf   TMR0
BP2_Loop
    comf   TMR0, w
    btfsc  STATUS, Z
    goto   BP_End
    btfsc  PORTB, 0
    goto   BP2_Loop

    goto   BP2

BP_End

    return

    end
```

When I went through this program the first time, I got the initial values listed in Table 8-1 for the file registers in a PIC16C84.

Table 8-1. First run results of the initial 16C84 file register values.

Address	Value
0x00C	0x0FF
0x00D	0x0FF
0x00E	0x0FF
0x00F	0x0BB
0x010	0x0FF
0x011	0x0FB
0x012	0x0FB
0x013	0x0FB
0x014	0x0FB
0x015	0x0FF
0x016	0x0FF
0x017	0x0F9
0x018	0x0FD
0x019	0x002
0x01A	0x0A1
0x01B	0x003
0x01C	0x070
0x01D	0x070
0x01E	0x0FF
0x01F	0x0FE
0x020	0x0FF
0x021	0x0FF
0x022	0x0FF
0x023	0x0FF
0x024	0x0FF
0x025	0x0FF
0x026	0x0F7
0x027	0x000
0x028	0x00A
0x029	0x00A
0x02A	0x008
0x02B	0x008
0x02C	0x003
0x02D	0x07F
0x02E	0x0FF
0x02F	0x0F7

I then powered down and waited a few minutes to let the PIC lose any internal values. When I read it again, I read the initial values listed in Table 8-2.

Table 8-2. Second run results of initial 16C84 file register values.

Address	Value
0x00C	0x0FF
0x00D*	0x0F7
0x00E	0x0FF
0x00F*	0x008
0x010	0x0FF
0x011	0x0FB
0x012	0x0FB
0x013*	0x010
0x014	0x0FB
0x015	0x0FF
0x016	0x0FF
0x017	0x0F9
0x018	0x0FD
0x019	0x002
0x01A	0x0A1
0x01B	0x003
0x01C	0x070
0x01D*	0x0FF
0x01E	0x0FF
0x01F	0x0FE
0x020	0x0FF
0x021	0x0FF
0x022	0x0FF
0x023	0x0FF
0x024	0x0FF
0x025	0x0FF
0x026	0x0F7
0x027	0x000
0x028	0x00A
0x029*	0x032
0x02A	0x008
0x02B	0x008
0x02C	0x003
0x02D	0x07F
0x02E*	0x000
0x02F	0x0F7

Note that most of them are the same value (I've just marked the ones with an asterisk that are different). This is because the flip-flops inside the PIC can power up to any value, but small variances in the parameters of the transistors that make up the flip-flops cause the flip-flops to initially power up to a certain state.

This program illustrates the importance of initializing file registers before using them and not relying on the fact that the simulators indicate that the power-up values are always 0x000.

This experiment could be repeated on the 16F84 and all 68 registers read out.

Reading the registers really isn't rocket science. I really think the most interesting aspect of the code is the switch debounce.

Typically, when a mechanical switch opens or closes, there is a very erratic signal output (Fig. 8-27) This bounce has to be filtered out; otherwise, there's a chance that multiple button presses will be read by the program, making its output unreadable by the user.

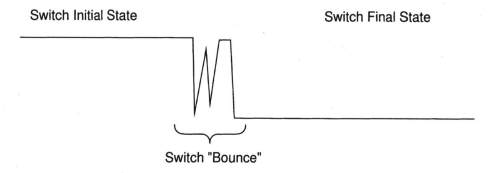

8-27 Switch bounce.

The "ButtonPress" routine waits for the PIC's input to be constant for at least 30 msec before accepting that it is in a valid state.

The pseudo-code for waiting for the button state to change reliably is:

```
BP                                ;  Wait for the button to go down
  Loop while ButtonHI

  reset TMR0                      ;  Timer0 is used to time button state
  Loop while Buttonlo & TMR0 < 30 msec

  if ButtonHI                     ;  Was there a bounce
    goto  BP                      ;  Yes
; The button has been down for 30 msec. Push down has been debounced
```

Typically 20 msec without a button input change is used as the debounce period. 30 msec was used in this program because it allowed easier TMR0 checking code (at 1 MHz and a 32x prescaler value, you will get a 33 msec delay).

Later, in PROG46, I will show how this debounce can be done using interrupts.

PROG47: Sleep

Putting the PIC to sleep can be somewhat confusing with respect to all the different ways it can be woken up. (See Fig. 8-28.)

8-28 PROG47 Experiment in sleep.

```
 title   "PROG47 - Experiment in sleep."
;
;  This program indicates the _TO and _PD bits when it wakes up
;    (along where the execution currently is).
;
;  Hardware notes:
;    Reset is tied directly to Vcc, and PWRT is enabled.
;    A 4.7K pull-up and switch pull-down are attached to PORTB.0
;    A 220-ohm resistor and LED are attached to all the PORTB.7:1
;    A 220-ohm resistor and LED are attached to all the PORTA.0
;
;  Myke Predko
;  97.02.22
;
 LIST P=16F84, F=INHX8M, R=DEC, N=45
 errorlevel 0,-305
 INCLUDE "\PIC\MPASM\p16c84.inc"


__CONFIG _CP_OFF & _WDT_OFF & _XT_OSC & _PWRTE_ON       ;  WDT off
;__CONFIG _CP_OFF & _WDT_ON & _XT_OSC & _PWRTE_ON        ;  WDT ON

   PAGE
;  Mainline of PROG47

   org     0

   movlw   0x0FD                    ;  Indicate that this is power-up
   movwf   PORTB

   bsf     STATUS, RP0
   clrf    TRISA & 0x07F            ;  Output _TO/_PD in PORTA
   movlw   0x0F9
   movwf   TRISB & 0x07F            ;  Output the Location through PORTB
   bcf     STATUS, RP0

Loop                               ;  Loop back here to display

   comf    STATUS, w                ;  Display _TO/_PD values
   movwf   PORTA

   bsf     INTCON, INTE             ;  Allow button press to wake up PIC

   sleep                            ;  Put the PIC to sleep
   nop

   movlw   0x0FB                    ;  Indicate that we are executing
   movwf   PORTB                    ;    after sleep

   goto    Loop                     ;  Loop around again

   end
```

Looking over the code in Fig. 8-28, you are probably wondering how something so simple can explain subtleties in the PIC, but it actually will show a lot.

For the first time you run it, both the _TO and _PD along with the RB1 LEDs will light. This is as expected; the PIC has powered up and gone through the reset address (0x0000).

Next, press Reset. The _PD LED will now go out. This is because you are resetting the PIC during "sleep."

Finally, press the input button at RB0. Now, the RB1 LED will go out and the RB2 LED will light. The _TO/_PD indicates that sleep was interrupted and the RB2 LED indicates that execution continued.

Now, let's modify the program a bit. First, comment out the "bsf INTCON, INTE" line and rerun the program. Start up the program and press the RB0 button.

What happens?

If nothing happens, keep pressing the button. Vary the length of time you push it down.

Sorry, I couldn't resist.

Nothing will happen because the RB0/Int hardware isn't enabled to wake the PIC up from sleep. Note that, if GIE was set (along with INTE), the interrupt handler would begin to execute (after the instruction after the "sleep" instruction). This execution of the instruction after the "sleep" instruction is why I always put a "nop" after "sleep" to make sure that nothing unexpected gets executed inadvertently.

Take that comment out, and now comment out the "_CONFIG" line commented "_WDT_OFF." Also take out the comment on the "_WDT_ON" line, and burn the program into the PIC.

Powering up, you'll see that the _TO, _PD, and RB1 LEDs will turn on (as expected). After waiting a few seconds, all three LEDs will go out and only the RB2 LED will go on. Again, this is as expected. Pressing the RB0 RST Button will get the same results as before.

Taking the chart explaining the _TO/_PD bits from the datasheet, I've enhanced it to reference to PROG47 (Table 8-3).

Table 8-3. The _TO/_PD bits

_TO	_PD	RB1	RB2	Description
1	1	1	0	Power-on reset/_MCLR during normal operation
1	0	1	0	_MCLR during sleep (pressing "RST")
1	0	0	1	RB0/int button pressed
0	0	0	1	WDT wake up from sleep

For PICs that have a TMR1, note that, if an external oscillator is used for it, it will run and wake up the PIC during sleep. TMR0 and TMR2 will not run during sleep.

I/O with interrupts

As I've said throughout the book, using interrupts can make a lot of your programming tasks much easier.

PROG45: Timer/interrupt handler with context saving

To try to demonstrate how simple interrupt handlers are, I've taken PROG17 and used an interrupt handler (using Timer0 as the source of the interrupts) to provide the delay. The timer and interrupt handler avoid the need to use a counter and loop around in the foreground waiting for the delay interval to pass. This means that while a hardware delay is taking place, other processing can be done.

The code shown in Fig. 8-29 differs in several areas; the most important in that all the LED handling is done in an interrupt handler and the delay is totally done in hardware (as opposed to the "delay" routine in PROG17).

8-29 PROG45 LED display in background.

```
 title  "PROG45 - LED display in background."
;
;  This program is a copy of Program 17, but the LED changing
;   happens in an interrupt handler using TMR0 for the delay
;   instead of an explicit delay loop.
;
;  This program is an enhancement to PROG16.ASM
;
;  Hardware notes:
;   Reset is tied directly to Vcc, and PWRT is enabled.
;   A 220-ohm resistor and LED are attached to all the PORTB bits
;
;  Myke Predko
;  97.02.20
;
  LIST P=16F84, F=INHX8M, R=DEC
  ERRORLEVEL 0,-305
  INCLUDE "\PIC\MPASM\p16c84.inc"

;  Registers
 CBLOCK 0x0C
Port                         ;  Value saved for the PORT
ONoff                        ;  Flag to indicate whether LEDs state
_w, _status                  ;  Interrupt context save values
 ENDC

  __CONFIG _CP_OFF & _WDT_OFF & _XT_OSC & _PWRTE_ON

  PAGE
;  Mainline of PROG45

  org   0

  goto  MainLine             ;  Jump to the mainline code

  org   4

Int                          ;  Interrupt handler

  movwf _w                   ;  Save the "w" register
```

8-29 PROG45 LED display in background. *Continued.*

```
        movf    STATUS, w        ;  Save the STATUS register
        movwf   _status

        bcf     INTCON, T0IF     ;  Turn off TMR0 overflow flag

        movf    ONoff            ;  Do we turn the LEDs on or off?
        btfss   STATUS, Z
        goto    Int_off          ;  They're on, turn them off

Int_ON                           ;  LEDs are off, turn them on

        btfsc   Port, 7          ;  Check to see if the high bit is set
        goto    Int_Set          ;   It is, start turning off the bits
        bsf     STATUS, C
        goto    Int_Rotate
Int_Set
        bcf     STATUS, C
Int_Rotate
        rlf     Port             ;  Now, shift the data over (change Carry)
        movf    Port, w          ;  Save the new value
        movwf   PORTB

        incf    ONoff            ;  Indicate that the LEDs are on

        goto    Int_End

Int_off                          ;  Turn off the LEDs

        movlw   0x0FF
        movwf   PORTB

        clrf    ONoff            ;  Indicate the LEDs are off

Int_End                          ;  Interrupt handler is finished, return

        movf    _status, w
        movwf   STATUS
        swapf   _w
        swapf   _w, w

        retfie


MainLine

        movlw   0x0FF
        movwf   PORTB            ;  Turn off all the LEDs
        movwf   Port

        bsf     STATUS, RP0
        clrf    TRISB & 0x07F    ;  Set all the PORTB bits to output
        ifndef  Debug
        movlw   0x0D7            ;  No Debug, Set OPTION to full wait
```

```
        else
          movlw   0x0D0                 ;   Debug, set OPTION to minimum wait
        endif
          movwf   OPTION_REG & 0x07F
          bcf     STATUS, RP0

          clrf    TMR0

          movlw   0x0A0                 ;   Enable the timer overflow interrupt
          movwf   INTCON

          clrf    ONoff                 ;   Initialize the interrupt flag to "off"


          goto    $                     ;   Could do anything from here...

        end
```

By putting the LED handling/displaying values in the interrupt handler, I have freed up the PIC mainline to do other things. Many high-end computer systems have a "Cylon Eye," which goes back and forth to indicate that the hardware is running. The code presented here could perform the same function. The interrupt handler handles the status output, while the mainline actually does the primary code.

Looking at the interrupt handler, you'll see that the context save and restore and interrupt hardware reset ("bcf INTCON, T0IF") are very straightforward and have been explained elsewhere in the book. The hardware reset is simply resetting the Interrupt Active flag before returning. If this flag wasn't reset, then the interrupt handler would immediately execute again (and again and again . . .) putting the program into an endless loop.

Looking over the code and comparing it to PROG17, I feel that PROG45 is more efficient. This is because the interrupt handler really doesn't intrude on the operation of the mainline—something that PROG17 didn't do.

Originally, when I was writing up this chapter, I created a program (PROG22) that was identical to PROG17, but it used TMR0 and an interrupt handler to create the delay turning on and off the LEDs (so a person can see it happening). If I compared this program to PROG17, I would say that PROG17 was more efficient, because it used less resources than PROG22. The lesson in this is don't use PIC resources just because they're "neat." Instead, using hardware resources in the PIC should be considered and the most efficient route taken.

PROG46: Debouncing inputs

PROG46 (Fig. 8-30) is a rewrite of PROG18 to use interrupts to provide a button-press debounce.

8-30 PROG46 Register contents in debounce.

```
  title   "PROG46 - Register contents in debounce."
;
;  This program reads the value in a RAM register and outputs it
;  inverted onto PORTB (which has LEDs to display the value). All the
;  RAM registers are read and displayed. A button is used as the
;  instigator of the next value read. The FSR is copied into the LEDs
;  to display the current register being displayed.
;
;  This program is a modification of PROG18.ASM to use the interrupt
;   handler to debounce the button input.
;
;  Hardware notes:
;   Reset is tied directly to Vcc, and PWRT is enabled.
;   A 4.7K pull-up and switch pull-down are attached to PORTB.0
;   A 220-ohm resistor and LED are attached to all the PORTB.7:1
;   A 220-ohm resistor and LED are attached to all the PORTA.0
;
;  Myke Predko
;  96.05.20
;
  LIST P=16F84, F=INHX8M, R=DEC, N=45
  errorlevel 0,-305
  INCLUDE "\pic\mpasm\p16C84.inc"

;  Registers
Reg             EQU 0x0C                        ;  Register to display


  __CONFIG _CP_OFF & _WDT_OFF & _XT_OSC & _PWRTE_ON

  PAGE
;  Mainline of PROG46

  org    0

  goto   MainLine

  org    4                        ;  Interrupt handler address
Int

  btfss  INTCON, T0IF             ;  Do we have a timer overflow?
   goto  Int_Switch               ;   No, handle the switch

  bcf    INTCON, T0IF             ;  Yes, turn off timer interrupts
  bcf    INTCON, T0IE

  bsf    STATUS, RP0              ;  Figure out if button release
  movlw  0x040                    ;  Flip INTEDG bit
  xorwf  OPTION_REG & 0x07F
  andwf  OPTION_REG & 0x07F, w    ;  Set Z flag if finished
  bcf    STATUS, RP0

  btfsc  STATUS, Z
```

```
        bcf     INTCON, INTE        ;   Turn off RB0 Int to indicate
                                    ;   button was pressed and
                                    ;   released

        goto    Int_End

Int_Switch                          ;   Interrupt on switch, reset timer

        bcf     INTCON, INTF        ;   Reset the interrupt flag

        clrf    TMR0                ;   Going to wait for key press
        bcf     INTCON, T0IF

        bsf     INTCON, T0IE        ;   Enable TMR0 interrupt

Int_End

        retfie


        PAGE
MainLine                            ;   Mainline of reading file regs

        movlw   0x0FF
        movwf   PORTB               ;   Turn off all the indicator LEDs
        movwf   PORTA

        bsf     STATUS, RP0
        movlw   0x01E               ;   Make PORTA.0 output
        movwf   TRISA & 0x07F
        movlw   1
        movwf   TRISB & 0x07F       ;   Set PORTB.7:1 bits to output
        movlw   0x094               ;   Set up prescaler for TMR0
        andwf   OPTION_REG & 0x07F
        bcf     STATUS, RP0

        movlw   Reg                 ;   Set up index to the start of regs
        movwf   FSR

Loop                                ;   Loop to here for each register

        comf    INDF, w             ;   Get the bits in "Reg" and display
        movwf   PORTB               ;   so the high bit LEDs are on
        movwf   PORTA

        incf    FSR                 ;   Point to the next register
                                    ;   read

        movf    FSR, w              ;   Are we at end of register space?
        addlw   0 - 0x030

        btfsc   STATUS, C           ;   If Carry is NOT set, then end
        goto    Finished

        call    ButtonPress
```

8-30 PROG46 Register contents in debounce. *Continued.*

```
    goto   Loop

Finished                              ;  Program is complete, loop forever
    goto   $

;  ButtonPress routine - Wait for the button to be pressed and released
ButtonPress

    movlw  0x090                      ;  Wait for KeyPress int to clear
    movwf  INTCON

    btfsc  INTCON, INTE
    goto   $ - 1

    bcf    INTCON, GIE                ;  Make sure all interrupts are off

    return

    end
```

Note that I don't bother to save the context registers ("w" and STATUS) in the interrupt handler because the check loop doesn't rely on either register. While this works in this case, it is *not* a recommended design practice.

Even though the debounce code has fewer instructions than PROG18's, it seems a lot more complex. To keep the code as small and as fast as possible, I've really made the interrupt handler a state Machine. When you first go through it, I'm sure it doesn't look very simple.

To make matters worse, the state machine uses hardware register bits (from INTCON and OPTION) to determine the next state, rather than a more traditional state value.

The state machine interrupt handler code can be represented with the following pseudo-code:

```
Int
   if Timeout                         ;  KeyPress, did it happen?
      Turn Off TMR0 Int               ;  No longer waiting for TMR0 int

      Toggle RB0/Int Going Up/Down    ;  Flip waiting for bit high/low

      if RB0/Int Going Down
         INTCON = 0                   ;  If int waiting for down, all done

   else                               ;  Key pressed

      Reset Timer and Enable Int      ;  Reset TMR0 counter and enable int

   retfie                            ;  Return from int
```

which should be quite simple to understand. The actual code is an example of how simple functions can be optimized.

This type of interrupt handler can be very desirable because of the speed it runs at. The timeout path, the longest, only takes 17 instruction cycles (including the interrupt branch) or 68 μsecs to execute.

Analog input/output

In the earlier chapters, I have described how some PICs have internal hardware for measuring and producing analog voltages. There are, however, a few tricks that you can do to provide analog I/O on a PIC that doesn't have any analog I/O.

PROG20: Measuring resistance values

Analog voltage input is the primary method of analog data input for the PIC. As noted earlier in the book, the 16C7x series of PICs have an internal ADC. For PICs without ADCs, this can be done by connecting an external ADC to the PIC.

If we were to look at how the analog voltage to be measured was produced, we would probably find that it was created by a potentiometer connected to a mechanical device that is set up as a voltage divider.

If this is the case, then the resistor value can be read easily by measuring the fall time in a simple RC Network as is done in this experiment (Fig. 8-31).

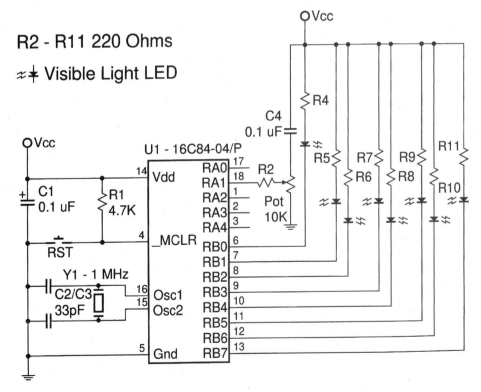

8-31 PROG20 schematic.

The PIC would first charge the capacitor to 5 V and then change the pin to input and wait for the capacitor to discharge through the resistor. If you were to put this on an oscilloscope, it would look like Fig. 8-32.

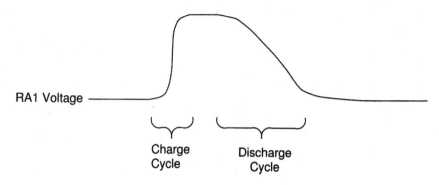

RA1 Voltage

Charge
Cycle

Discharge
Cycle

8-32 A/D charge/discharge cycle.

From basic electronic theory, we know that the time required for the capacitor to charge is:

$$time = R \times C \times \ln\left(\frac{Vend}{Vstart}\right)$$

where *Vstart* and *Vend* are the starting and ending voltages that we are interested in. For the PIC, we would be interested the capacitor voltage starting at Vdd (after being charged by the PIC to 5 V) and then waiting for the capacitor to discharge to the input transition point (1.5 V in the PIC).

Because we know the capacitor value along with the voltages and the time it took for the capacitor to discharge, we can rearrange the previous formula to find *R*:

$$R = time / \left(C \times \ln\left(\frac{Vend}{Vstart}\right)\right)$$

Therefore, by controlling the voltage applied to the network and knowing the value of the Cap, we can determine the value of the resistor.

The method described here is actually the method used by the Parallax Basic Stamp II to test a resistor value. It is actually quite an elegant solution to the problem of reading a resister value.

The code used to test the analog I/O uses the following logic:

```
Initialize I/O ports and timer
while 1 == 1                    ;  Loop forever
  Charge the cap                ;  Set up cap for the resistance measure
  clear the timer               ;  Time cap to discharging through resistor
  Set pin to input              ;  Allow the cap to discharge
  while pin == 1                ;  Wait for the cap to discharge
  Display Timer Value           ;  Cap discharged, display the timer
```

This code is unique in that no RAM registers are used for the timing. This is totally done within the PIC hardware. The timer does not have to be used, instead a simple timer could be used.

The actual PIC Assembly source is shown in Fig. 8-33.

8-33 PROG20 Reading a resistor value.

```
title   "PROG20 - Reading a resistor value."
;
;   This program copies the "RCTIME" instruction of the Parallax Stamp.
;   A resistor value is read repeatedly and displayed.
;
;   This program is a modification of PROG17.ASM
;
;   Hardware notes:
;   Reset is tied directly to Vcc, and PWRT is enabled.
;   A 10K pot along with a 0.1-uF cap and 100-ohm series resistor
;     on PORTA.0
;   A 220-ohm resistor and LED are attached to all the PORTB.7:0
;
;   Myke Predko
;   96.06.02
;
    LIST P=16F84, R=DEC
    ERRORLEVEL 0,-305
    INCLUDE "\PIC\MPASM\p16c84.inc"

;   Registers

__CONFIG _CP_OFF & _WDT_OFF & _XT_OSC & _PWRTE_ON

    PAGE
;   Mainline of PROG20

    org     0

    movlw   0x0FF
    movwf   PORTB               ;   Turn off all the LEDs
    clrf    PORTA               ;   Use PORTA as a counter

    bsf     STATUS, RP0
    clrf    TRISB & 0x07F       ;   Set all the PORTB bits to output
    movlw   0x0D0               ;   Set up the timer to fast count
    movwf   OPTION_REG & 0x07F
    bcf     STATUS, RP0

    movlw   TRISA               ;   Have to set/read PORTA.4
    movwf   FSR

Loop

    bsf     PORTA, 0            ;   Charge cap on PORTA.0
    bcf     INDF, 0             ;   Make PORTA.0 an output
    clrf    TMR0                ;   Now, wait for the cap to charge
Sub_Loop1                       ;   Wait for the timer to reach 10
    movf    TMR0, w             ;   Get the timer value
    sublw   10                  ;   Is it greater than 10?
    btfsc   STATUS, C           ;   Is the Carry flag set?
     goto   Sub_Loop1           ;   Yes, loop around again

    bsf     INDF, 0             ;   Now, wait for the cap to discharge
```

8-33 PROG20 Reading a resistor value. *Continued.*

```
 clrf    TMR0                     ;   and time it.
Sub_Loop2                         ;   Just wait for PORTA.1 to go low
  btfsc  PORTA, 0
  goto   Sub_Loop2

  comf   TMR0, w                  ;   Get the timer value
  movwf  PORTB

  goto   Loop                     ;   Get another time sample

  end
```

As you run this experiment, you'll notice how dependent the circuit is on the capacitor value. After you get this circuit running, you should try changing the capacitor with different ones of the "same" (marked) value. You'll find that you will get a real variance in the value measured in the least significant bits.

Because of the variance to the capacitor value, I would not recommend this circuit for critical resistance measurements. Yes, a precision cap and power supply, along with characterizing the timer values from the PIC, would give accurate results; however, like using a precision RC clock, this is not reasonable for volume production. If accuracy better than the most significant 5 bits of the timer are required, I would recommend using an ADC attached to the PIC (or a 16C7x, which has built-in ADCs), rather than this circuit.

The advantages of this circuit and software are its simplicity and the few PIC resources used. While not providing high accuracy, the circuit does provide excellent repeatability that can be very useful in many applications.

One last thing to notice about this circuit is the use of FSR to point to TRISA (for the switching between output and input in the A/D function). FSR is loaded with the address of TRISA (0x085), and using INDF, TRISA is accessed directly without having to change the default bank (the usual "bsf STATUS, RP0" instruction).

PROG31: Resistor ladder output

While a single reference voltage might be useful for some applications, a variable voltage output is much more useful for many other applications.

Outputting analog voltages directly from the PIC is actually quite simple. This example program and hardware will show you how to output a "Sawtooth wave" with 0.55-V increments.

The output voltage is determined from a voltage divider, which has the formula:

$$\text{Vout} = \text{Vcc} \times \left(\frac{Rn}{(Rs + Rn)} \right)$$

A "variable resistance" voltage divider can be implemented on the PIC as a "resistor ladder" like the VRef circuit of the 16C62x devices.

The PIC analog voltage output works by the principle of being able to vary the "lower resistance" of the voltage divider. In doing this, the "lower resistance" (labelled Rn in the previous formula) is changed by selecting a different ground point within the ladder. Earlier in this book, I showed how a standard I/O bit could be used as an open-collector output. This was done by loading a bit with zero and then enabling the bit to output to pull the line down to approximately ground potential.

In Fig. 8-34, you can see that, by enabling an output bit and setting the output low, the voltage divider circuit has its bottom (or ground). By changing the bit, this causes a change in the output voltage. The resistance values shown in Fig. 8-31 are incorrect for this experiment (this circuit was used for trying out enabling parallel resistances).

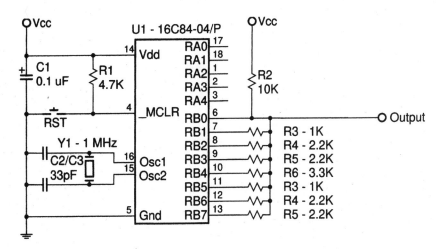

8-34 "PROG31" schematic.

To calculate the resistance values, I've rearranged the previous formula to:

$$Rn = \left(\frac{\text{Vout}}{\text{Vcc}}\right) \times \frac{Rs}{\left(1 - \left(\frac{\text{Vout}}{\text{Vcc}}\right)\right)}$$

By choosing a value for *Rs* (say 10K), we can easily calculate the value of *Rn* for a given Vout.

The reason why I am going to all this trouble to come up with a formula and calculate it out is to ensure that I can get reasonable linearity in Vout for different bit outputs. With the resistor ladder connected to port "B", nine different voltages can be output. In selecting these voltages, I have tried to space them evenly.

With the circuit shown in Fig. 8-34, you can get nine different voltages very simply. The first is when all the output bits are turned off (which leaves the largest *Rn* possible). At this point, the Vout will be at its maximum value. Set Vout equal to zero volts is accomplished by outputting a zero on bit zero (pulling the output to ground level). The remaining seven voltages are selected by grounding intermediate resistors in the ladder.

One *very* important thing to keep in mind is that a resistor ladder such as this is very poor at maintaining the output voltage if it has to source or drive current. If there is any type of tangible current flow (more than a few microamps), outside of the voltage divider, the output voltage will be changed from what you expect. The best way to avoid this is to buffer the resistor ladder output.

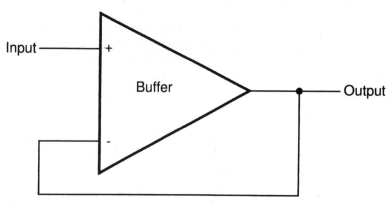

8-35 Unity gain voltage buffer.

Now, the same value could be used for each resistor in the ladder. However, if we were to plug these into the PIC, we would find that we would get the voltage output as shown in Fig. 8-36.

This output obviously deviates significantly from the desired (linear) output. Actually, finding the correct resistor values is not that difficult. Using the rearranged formula for Rn (given previously), we can plug in an Rs value of 10K and a Vcc of 5 V, then figure how to go from 0 to Vmax in nine steps.

The values in PROG31 use standard resistor values to try to get as linear output as possible.

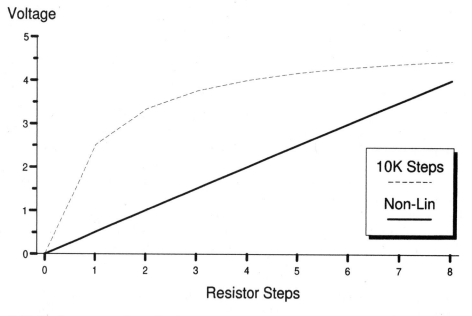

8-36 Nonlinear versus linear R values.

Schematically, we end up with the circuit shown earlier in Fig. 8-34.

The reset, _MCLR, and power are the standard values from the previous PIC examples you have been working with. One of the really nice things about this circuit is that it can be checked for wiring errors without the PIC installed. This is done by hooking up your Dmm between Ground and Vout. Power up and then short various resistor interconnects to ground. You will see the voltage output change to values very close to what I specified in the table above.

If you're curious, you could rewire the circuit using a constant value for each step in the ladder. If you repeat the experiment with the voltmeter, you will find the asymptotic curve predicted in Fig. 8-35.

The PIC program itself is very simple. Along with outputting the sawtooth, I toggle an LED to show that the PIC is running and when the value is changing.

The pseudo-code for the program is:

```
setup port "B"                 - Want to drive R-Ladder, so clear all bits
LSB_rotate = 0                 - Want to shift in a low to the output
                                 control

while 1 = = 1                  - Loop forever

  toggle LED                   - Indicate the program is running

  shift the Open Collector O/P - Change the output value.
```

In the actual code (Fig. 8-37), I cheat a bit to make it simpler to program. I make sure that I don't change the Carry flag (called the "LSB_rotate" in the pseudo-code).

8-37 PROG31 Resistor/adder analog output.

```
title  "PROG31 - Resistor ladder analog output."
;
;  This program runs through a sawtooth analog output from the
;  16C84. The output is generated by a resistor ladder attached
;  to PORTB. To set a particular voltage, a bit is output to 0 V.
;
;  Hardware notes:
;   Reset is tied directly to Vcc, and PWRT is enabled.
;   The resistor ladder is attached to PORTB.7:0
;
;   A 10K resistor between PORTB.0 and Vcc (output is taken from
;     here as well)
;   A 1K resistor between PORTB.0 and PORTB.1
;   A 2K resistor between PORTB.1 and PORTB.2
;   A 2K resistor between PORTB.2 and PORTB.3
;   A 3.3K resistor between PORTB.3 and PORTB.4
;   A 3.3K resistor between PORTB.4 and PORTB.5
;   A 3.3K + 4.7K resistor between PORTB.5 and PORTB.6
;   A 10K + 4.7K resistor between PORTB.6 and PORTB.7
;   A 47K resistor between PORTB.7 and Ground (top level)
;
;   Myke Predko
;   96.06.27
;
```

8-37 PROG31 Resistor/adder analog output. *Continued.*

```
LIST P=16F84, R=DEC
errorlevel 0,-305
INCLUDE "\PIC\MPASM\p16c84.inc"

;  Registers
Count            EQU 0x00C
Counthi          EQU 0x00D

__CONFIG _CP_OFF & _XT_OSC & _PWRTE_ON  & _WDT_OFF

PAGE
;  Code for PROG31

   org   0

MainLine

   clrf    PORTB              ; Output only lows from PORTB

   movlw   TRISB             ; Set up the TRIS values
   movwf   FSR

   bcf     STATUS, C         ; Use the Carry as the skip value

Loop                         ; Loop around here to output the sawtooth

   ifdef   DMM               ; Just Dlay if only a Dmm available for
   call    Dlay              ;   seeing the output
   endif

   rlf     INDF              ; Shift over the output bit

   goto    Loop

   ifdef   DMM
Delay

   movlw   194               ; Now, display for a half a second
   movwf   Count
   movlw   163
   movwf   Counthi
Dlay
   decfsz  Count
   goto    Dlay
   decfsz  Counthi
   goto    Dlay

   return
   endif

   end
```

One thing to note in the code is that it is designed to be used with an oscilloscope for output (see the scope picture in Fig. 8-38). If this is not available, you can define "DMM" and a half second delay is put in so the output voltages can be seen on a slow device like a DMM.

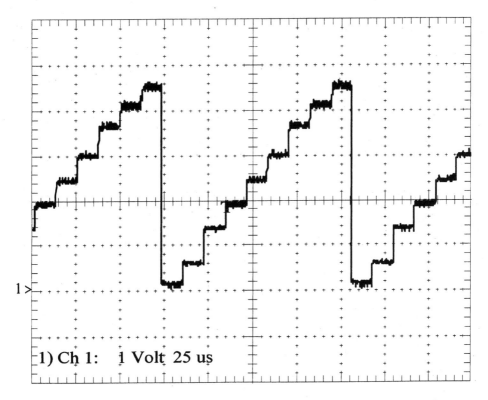

8-38 Actual VLadder output.

As you step through the code, you will watch the zero shift through TRISB, which causes PORTB to act like a programmable open-collector output and changes the effective voltage divider. When you look at a scope picture that I've done (Fig. 8-38), you'll notice that the output voltage doesn't reach a full 5 V. This is because there's always a resistor value between Vcc, Output, and Ground. To get the full 5 V, RB0 (the bottom value) would have to output Vdd, rather than just zero.

As well, you'll notice that some steps are larger than others. This is because of the problems trying to match standard resistor values to the calculated ratios. If I had been willing to play around with values a bit more, I probably could have gotten a sawtooth wave where every step was exactly matched.

Resistor ladder outputs can be used for applications other than a wave-form generator. One of the most personally intriguing applications that I've always wanted to try was to use the output to generate a composite video signal.

Closing out the experiments

If you've made it here and have worked through each experiment, you're probably thinking you know everything there is to know about the PIC.

I've probably gone through 25% of what there is to know about creating PIC software and hardware applications.

However, through all this, you should have gotten a good idea of what the PIC is capable of in just a few lines of code. None of the experiments in this chapter use more than 50 instructions (not counting tables), and each does some pretty significant things.

In the next chapter, I'll go through some applications and show you what the PIC is really capable of.

9

CHAPTER

Projects

The whole point of working with the PIC is being able to develop useful things with it. I've included a number of applications to show you how applications are developed. I've tried to make the projects presented here very accessible, with as few difficult-to-find parts as possible. In the few cases where this wasn't possible, I have identified the source for part and have provided ordering information in appendix D.

When I was doing my final proofreading for the book, I felt that the projects would be best used to demonstrate different aspects of the PIC. For this reason, I have not created raw card designs for the projects presented in this chapter. Obviously, this is not true for the programmer and emulator that I have designed for the book, and this is why I have included raw card images. The purpose of the book is to provide you with knowledge for developing your own applications. By not providing complete assembly directions on every project, I'm hoping you will gain skill in creating, debugging, and modifying the projects for your own use and preferences.

When I talk about parts and availability, for the most part, the parts used for the applications are available from such full-service sources as Digi-Key. Some other parts have been picked up at surplus stores, and you might not be able to find equivalents. In these cases, I've based the project description on how the parts were used and how you can replicate their functions. For example, for reading the keyboard, I picked up a surplus keyboard for $1, but the keyboard project isn't so much about interfacing to a *specific* keyboard, but to any switch matrix keyboard.

The purpose of the projects is to provide instruction both in how the PIC works and in how different interfacing hardware works. In going through this chapter, you will end up with blocks of code and hardware interfaces for your own projects.

Please note that the projects presented here are copywritten, and while I want you to go ahead with using the applications, I am not releasing the applications for you to profit from (either as kits or finished products).

Many of these applications were developed on the 16C84. While there's no reason why they would not work on a 16F84, you should simulate the code before burning it into a device to make sure it works as expected.

Blinking lights for Christmas decorations

Every fall, if you look through the hobbyist magazines, you'll see a number of projects that use a PIC to provide a decorative light display using a PIC and some LEDs.

Not being all that original, one of the first projects that I did on the PIC was a Christmas tree with flashing LED lights. A year later, not happy with what I did, I created a "Frosty the Snowman" that had flashing LEDs in his scarf and hat and played "Frosty the Snowman" virtually incessantly (I did put in a switch to turn off the music).

The code and hardware that I created to play the music (which is done in the background) are presented in appendix C, and the code is available on the diskette that comes with the book.

However, this still leaves how the LEDs were controlled.

When I was creating this chapter, I looked back at the two projects (and what I had written up to this point) and noticed a couple of things.

The first being that, if you don't know how to get a PIC running and turning on and off LEDs by now, you should go through the previous chapter again. The other was that the music software is presented in appendix C.

So, with these aspects already well discussed, what could I discuss that would be unique for this chapter?

What I haven't described well is how the "random" pattern of LEDs lighting/turning off was done. In both cases, a Linear Feedback Shift register (LFSR) was used to create the (pseudo) random pattern that the LED's display.

LFSRs are marvelously simple devices for creating pseudo-random patterns from register contents. These patterns are used for a variety of purposes. The "classic" one is generating a CRC (Cyclical Redundancy Check) code for use in communications. Passing a message through a LFSR will result in a series of bits unique for the message that can be compared at the receiver.

A standard LFSR (and used for both projects) is shown in Fig. 9-1.

The Linear Feedback Shift Register used in Fig. 9-1 takes the output of each bit of the shift register for the LEDs, and the final value is fed back into the input of the LFSR. This creates roughly 61,000 different patterns (for 16 bits), which, if each pattern is displayed a half a second apart, means that the data will not be repeated for almost eight and a half hours.

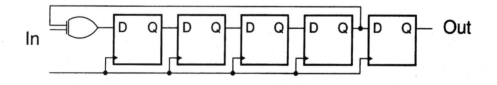

"D" Flip Flops Arranged as Shift Register

9-1 Example single "Tap" :FSR.

The LED control code could be described as:

```
Turn all LEDs on        ; Make sure all the LEDs work
Dlay500ms
Turn all LEDs off       ; Initial condition is that LEDs are off.
Loop
Dlay500ms
Get next LED value      ; Figure out the next value for the LEDs
Display new LED value
goto Loop
```

In the Christmas tree lights project, the calculation for the new LED value was done entirely in software using the following code:

```
bcf       STATUS, C     ; Start with putting in 0 LFSR shifted
btfsc     LFSRHI, 7     ; If high bit is set, Carry set
 bsf      STATUS, C
clrw                    ; Now, XOR Carry bit (high bit) with bit
btfsc     LFSRHI, 5     ; 13 for Tap
 movlw    1
xorwf     STATUS        ; Carry has LFSRHI.7 ^ LFSRHI.5
rlf       LFSRlo        ; Shift the data up with LSB
rlf       LFSRHI        ;  LFSRHI.7 ^ LFSRHI.5
```

I'm fibbing; the code used was considerably more complex than this. However, when I was writing this, I saw how it could be improved.

The LFSR register values were then output on a simple parallel bus using the circuit shown in Fig. 9-2.

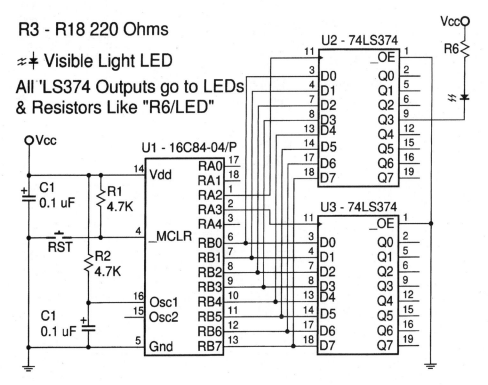

9-2 Christmas tree connection.

This worked quite well, but I wasn't happy with it, namely from the aspect of the amount of processing the PIC did and the amount of wiring I had to do (the first Christmas tree was all hand wired).

So, I got the bright idea of combining the PIC hardware with the external LED-driving hardware and came up with the circuit shown in Fig. 9-3.

In this circuit, the PIC simply shifts out the XORed output and input, Tap bit. Note that the schematic has been simplified. In both cases, not all the resistor/LED pairs have been shown (and the output of each '574 bit is put into the input of the next bit). For the "Frosty" schematic, I have not included all the shift loops, instead, on U3, I have referenced where each output goes.

The code to do this is:

```
bcf        PORTB, Out      ;  Start with the output bit low
btfsc      PORTB, End      ;  If End is high, then start with high
 bsf       PORTB, Out
clrw                       ;  Now, XOR the End with the Tap
btfsc      PORTB, Tap
 movlw     1 << Out        ;  Create the XOR value

xorwf      PORTB           ;  Get the correct output value

bsf        PORTB, Clk      ;  Clock out the new output bit
bcf        PORTB, Clk
```

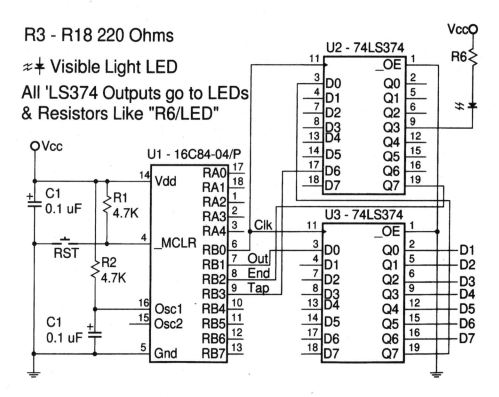

9-3 "Frosty the Snowman" connection.

By combining the external shift register hardware, I was able to come up with a very neat circuit that only required two more wires than a straight shift register.

Now, you might be asking yourself, why didn't I just use the Christmas tree code to create the random values and just shift them out in the "Frosty" project? The reason for that was because I was afraid of the LEDs "flashing" when the shift register was updated; after I finished the Christmas tree, I decided to make the LFSR 24-bits long and simply had the PIC drive the extra eight bits all the time (hoping that the flashing when the other bits '374s were updated wasn't too bad).

As I've said elsewhere in this book, the two Christmas projects were used to debug a compiler that I was writing. When I put the additional LEDs on the Christmas tree, there was quite a bit of noticeable flashing (even though I was running with a 500-KHz R/C clock).

To avoid this, I went with the combined PIC/'374 circuit.

For both of these projects, I used the circuit shown in Fig. 9-4 to provide +5 Volts (Vcc/Vdd).

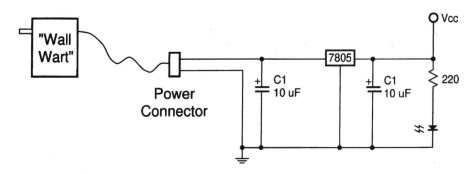

9-4 +5-V power for ac.

The "wall wart" is an ac-to-dc adapter that you can buy almost anywhere for just a few dollars. Unless batteries are used, this circuit is used in each project shown in this book. Using a "wall wart" is a safe, cost-effective way of providing power to your projects.

You should also notice that I have put in an LED and resistor to indicate when power is active in the circuit. This is also very important and will eliminate your first question when a project doesn't work: "Is it getting power?" A switch is optional and, if desired, should be put between the power connector and "C1" on the positive voltage.

I realize that, for the Christmas projects, I really haven't given you a project to build. Again, the reason for this is because I believe that I have explained how to do most of the projects elsewhere in the book. The real magic of the Christmas projects is in creating a twinkling, seemingly random light show, and that was described in this section.

PROG35: Serial-LCD interface

One of the true last arts of computer electronics is figuring out how an RS-232 line should be set up. This project (Fig. 9-5) was originally designed to be an elec-

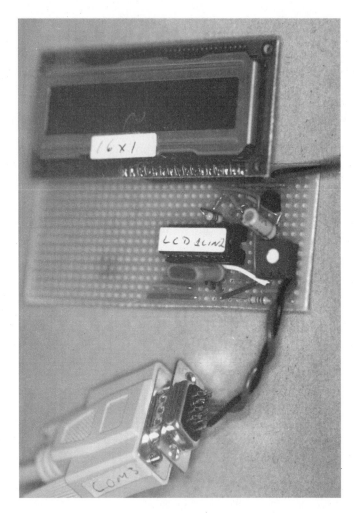

9-5
Serial-LCD interface.

tronic "break-out box" because of a number of problems people at work had with setting up "standard" RS-232.

The original project and hardware became very complex (provide a bridge between two communicating RS-232 devices) and really was too much for a small PIC. It also required a very complex user interface. However, as I was fooling around with a Liquid Crystal Display (LCD) module, I discovered how useful they could be. From this work, I really became interested in the idea of developing a serial-LCD interface. The project presented here is a result of this work and is really scaled down from the original project (originally, showing data being transmitted between two RS-232 devices).

There are other devices on the market that I could have bought, but there are aspects of the devices that I didn't like. These aspects centered around the cost of the devices, the need to set external jumpers for the type of signal coming in (RS-232 or TTL/CMOS digital), and its speed. The jumpers really limited the usefulness of the devices to me (I really hate having to set jumpers and, even worse, keeping documentation around for a little tool).

Rather than buying something, I wanted basically a device that would be "Plug 'n Play"—just attach it into a circuit on a serial line and let it figure out the type of data and its speed. Along with being able to figure out what speed the incoming data was running at, I also wanted to be able to handle extremely fast speeds. This would minimize the amount of code space and file register overhead required from the "host" computer (i.e., a PIC project) and minimize the delay required to send data.

The period of each bit is expressed as the baud rate and is in bits per second (bps). I tend to use the two terms interchangeably (I know they aren't exactly the same thing, but in common usage, they are used interchangeably). As can be seen back in Fig. 3-27, the actual byte data rate is at least 10 times less than the baud rate.

Just to show how extreme the performance that I was looking for was, I used the following snippet of code as a design point:

```
SerialOut                              ;  Send the byte in "w" out serially
    movwf       Temp                   ;  Save the byte for Shifting

    bcf         PORTA, w               ;  Send out the serial start bit
    rrf         Temp                   ;  Put least significant bit in Carry
    rlf         PORTA, w               ;  Shift out the Carry (Bit 0)
    rrf         Temp                   ;  Now, send out bit 1
    rlf         PORTA, w

      .
      :
    rrf         Temp                   ;  Shift out bit 7
    rlf         PORTA, w
    nop
    bsf         PORTA, 0               ;  Send out the stop bit


    return
```

This means that data would be sent at ⅛ the processor clock speed (it can be sent faster by using built-in hardware of some PICs). With a 1-MHz PIC, this would translate to 125 Kbps.

This was my design point. Once I had this requirement, I began to look at hardware.

Because I wanted to support arbitrary data speeds, using PICs with built-in serial hardware would not be an option. I would have to develop routines for reading the data rate and converting it to delay values that could be used in the data read.

To get the data speed, the bit length is measured on the product. I chose the carriage-return character because the first three bits alternate and are of a constant length.

To measure the time between transitions, I used the simple timer code:

```
btfsc       PORTA, 0               ;  Wait for the line to go low
    goto        $ - 1
    incfsz      Count                 ;  Timing loop
    incf        Counthi
    btfss       PORTA, 0
    goto        $ - 3
    movf        Counthi, w            ;  Counthi = Count - Counthi
    subwf       Count, w
    movwf       Counthi
```

This gives a timing granularity of 5 instruction cycles (20 clock cycles). This means that the bit could change anytime within those five instruction cycles. With this code, it also meant that the count is going to be high (by up to 5 cycles). This error rate is what I wanted to watch out for.

If the first bit is polled at some point in its "window," the next bit will be polled at that point plus the error. Over the 8 bits, this error is multiplied. For example, if the error was 7.5% per bit, over 8 bits, the cumulative error would be 60%.

This means that, if the polling took place more than 40% from the start of the first bit, then the last bit would be missed altogether!

To prevent this, I felt that the design should never have any possible bit error greater than 5% and that the bit sampling had to start in the first 25% of the start bit. This seemed like a reasonable margin and would not be an issue for lower speeds—just higher ones (like 125 Kbps).

To maximize the data rate that the receiving PIC could handle, I obviously had to maximize the speed in which the receiver ran at. 20 MHz was the maximum speed available in low-end and mid-range PICs at the time I did this (actually, at time of writing, only the high-end PICs are capable of speeds greater than 20 MHz). Running with a 20-MHz clock meant that I would have an instruction cycle time of 0.2 μsecs, and the timing loop would have a granularity of 1 μsec.

At 125 Kbps, a 1-μsec error works out to a potential error of 12.5% per bit. This is obviously unacceptable.

So, for very fast data streams, I use the TMR0 (remembering that the carriage return's first three bits alternate) module available on all PICs setup to run off of the instruction clock (incrementing TMR0 once every two instruction cycles) along with the polling loop:

```
btfsc       PORTA, 0        ;  Wait for the bit to start
 goto       $ - 1
clrf        TMR0            ;  Reset the timer
btfss       PORTA, 0        ;  Wait for the bit to end
 goto       $ - 1
incf        TMR0, w         ;  Add one to timer for initial loop
movwf       Count           ;    and clear instruction
```

This code gives the count accurate to two clock cycles (or at 20 MHz, 0.4 μsec). This works out to a maximum 5% error rate at 125 Kbps, which is the target.

The next issue to be resolved was how to poll and confirm the start bit (within 25% of the start bit). To confirm that there actually is a valid start bit, the line is checked one-eighth of a bit period after the start has been encountered. This reduced the amount of time from the transition the first bit could be checked to 12.5% (or one-eighth of a bit period). Doing the math, this meant that a maximum of 5 instruction cycles (at 20 MHz) could run between the start of the data and when the code started reading it.

The original solution to this was to create a giant state table and return every few instructions to poll the data input line. (This is "LCD1LINE.ASM" on the disk.) Complicating this was the code used to poll the line. Because I wanted to be able to handle both TTL/CMOS and RS-232 level signals, I used the following code:

```
movwf       _w              ;  Save next table execution point

movf        PORTA, w        ;  Get the input bit
xorwf       Polarity, w     ;  Confirm that it has changed
andlw       1               ;  Just isolate the bit
```

```
btfss           STATUS, Z       ; If zero, no transition
   call         ReadRS232       ;  Else, read the incoming

movf            _w, w           ; Restore the table index

addwf           PCL             ; Jump to the table entry

goto            ...             ; Table of gotos
```

This bunch of code itself could burn up as much as 17 instruction cycles before calling "ReadRS232" *without* even executing any useful code.

The final solution was to make the line change polling part of the ReadRS232 routine and have it called every few instructions (to a maximum of five). This method eliminated the overhead of the table jumps ("addwf PCL/goto ...") along with the need to save "w." While it did not meet the target of 5 instruction cycles, it was a vast improvement over the earlier code. When I ran the code, I noticed an immediate improvement in the validity of the read-in data.

LCDs have a reputation for being difficult to work with. Before this project, I did a fair amount of research to understand what was the correct type to get and how to interface to it. What I discovered was that, for ASCII displays, most use the Hitachi 44780 interface chip. It appeared that this would simplify the amount of work that I had to do on the project significantly.

Unfortunately, I ended up spending a couple of days on the Internet trying to find a good datasheet for the 44780. I found five different FAQs and datasheets, each with different errors and inconsistencies (including some LCD manufacturer's datasheets). After spending a day or so figuring out the discrepancies, I created "LCDFTEST.ASM," which took data from a PC at 9600 bps and displayed it on the LCD display. I was gratified to discover that this worked perfectly right from the start.

LCDs (especially with the Hitachi interface) are not difficult to use; they just require that the correct data be sent to them. For this reason, I suggest that you look over the code contained within this and the next section.

With the LCD, I had the two basic parts to my project, which now had to be merged. Because there are no 16C5*x* parts that contain EEPROM, and I don't have an EPROM eraser at home, I decided to debug the application on a 16C84. I thought this would be a fairly easy transition.

I developed the circuit shown in Fig. 9-6, which could be used interchangeably by the 16C54 and 16C84 (the only difference would be the maximum crystal frequency the project would work at).

Electrically, I was right on the money. Unfortunately, this doesn't translate well from the code side.

Looking back, I can summarize the problems:

- In the 16C5*x* family, subroutines have to take place in the first 256 addresses of a 512-address page.
- There are only two levels of subroutines you can nest. In the 16C*xx* parts, you have eight levels.
- There are no "addlw"/"sublw" instructions, which means that values have to be put in temporary registers, taking up file register space and code address space.

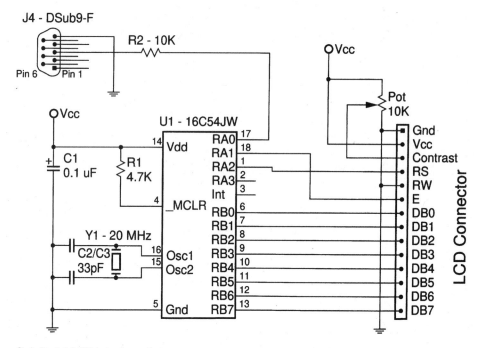

9-6 Serial-LCD interface diagram.

- There is no interrupt capability on the 16C5x. If interrupts could be used, the 5 instruction start check limit could be observed.
- There are only 512 addresses in the 16C54, as opposed to 1024 in the 16C84.
- The top 3 bits of the FSR register are always set to 1 in the 16C5x. When you go through the code, you will see how I had to compensate for this.

I originally chose the 16C54 because of its ability to run at high speeds (20 MHz) and its low cost. Looking back, I realize that this was the wrong choice for this application. The problems listed here made porting the code from the 16C84 very challenging. If I were to approach a project like this in the future, I would consider using a 16C61 because of its close relationship with the 16C84 (basically the same part, but with EPROM program store and no EEPROM memory).

The resulting program ("LCD1LIN2.ASM") probably looks like a mess. This is because I had to relocate all the subroutines to the first 256 address block of the device (instead of putting them, as is my custom, at the end of the program). I also had to put in numerous areas of conditional code, which also contributed to the general "messiness" of the code.

I have not included a listing of this code in the book because of its length (1000 lines) and difficulty to understand.

The biggest savings that I can claim to this project is the amount of simulation that I did before burning the 16C54. (The final simulation file "LCD1LINE.STI" is included.) This file simulates sending out a number of different serial characters for a number of different circumstances (e.g., new lines, carriage returns [screen clears], and backspaces), and I reran it numerous times, making sure that it worked perfectly for all different cases.

The payoff was that the program worked perfectly the first time that I burned it into a 16C54 and tested it. This is the code I've included here.

The serial-LCD interface is going to become a staple debugging tool for me in the future. I actually considered including its use in chapter 8 but declined because of the advanced code required to interface with it. The ease in which data can be sent can make debugging your program significantly easier. A more advanced version of this project (known as the SLI) is available from Wire Electronics (address in Appendix D).

PROG25: I/R tank

Robots are something that most engineers enjoy playing around with. I'm sure if you were to ask what is the appeal, you'd hear a lot of comments about their interest in: digital-control theory, artificial intelligence, and electro-mechanical interfaces.

Personally, I think most robots are built so their creators have something to boss around.

In this section, I will present two different robots that can be controlled by a PIC. This section will show you how to use the PIC and a TV remote control to send commands to a tracked vehicle. (See Fig. 9-7.) At the end of the chapter, I'll go through a programmable servo controller.

The genesis of this project was an article in *Electronics Now*. (Earlier in this chapter, I made the comment about how hobbyist magazines always have Christmas decorations in their November/December issues. If you watch them, you'll notice a number of different robot designs throughout the year.) After looking through the

9-7 The I/R tank.

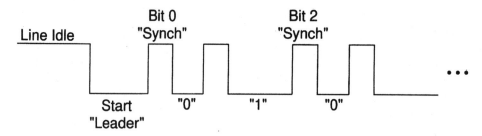

9-8 I/R TV remote data stream.

article, I decided to build my own robot, but there were a few things that I could improve upon.

The robot in the article used a single infra-red (I/R) receiver (typically used with TV remote controls) to control the robot. Two motors were controlled (each backwards and forwards) by an "H" bridge motor control made out of discrete transistors.

The magazine robot used a PIC 16C5*x* for control. I felt that, by using a mid-range part, I would have the advantage of interrupts to handle the incoming commands (which I'll refer to as *data packets* in this section and the next).

Most (if not all) I/R TV remotes use a Manchester encoding scheme in which the data bits are embedded in the packet by varying the lengths of certain data levels. (See Fig. 9-8.)

The normal signal coming from an I/R receiver circuit is high when nothing is coming (line idle) and then goes low with a "leader" signal to indicate that data is coming in. The data consists of a bit "Synch" that, when it completes, the bit value is transmitted as the length of time before the next bit "Synch."

So, to read these values, I used the code shown in Fig. 9-9 as an interrupt handler. This code has the advantage of being able to discriminate between different manufacturer's I/R codes (the differences lay in the length of time during the transitions, the number of bits, and the I/R carrier frequency).

9-9 An interrupt handler to read the values.

```
Int

   if T0IF                         ;   TimeOut—Invalid data being received
     Reset BitCount                ;   Reset everything to receive the
                                   ;    next packet
     Reset State
     Set INTCON to Just Int
   else
     switch State                  ;   Handle the incoming data
        0:                         ;   leader coming in
           Setup Leader TimeOut
           State = State + 1
           Value = 0               ;   Haven't received anything yet
```

```
    1:                              ; End of leader
        Setup Leader/Synch TimeOut
        Setup Bit Synch Int
        BitCount = 12
        State = State + 1

    2:                              ; Start of synch bit—come back for
                                    ;   each one
        Setup DataBit TimeOut
        State = State + 1

    3:                              ; Bit ended—use TMR0 to get value
        Value = Value >> 1
        If TMR0 > Threshold
            Value[lsb] = 1          ; TMR0 is more than "0" threshold
        BitCount = BitCount - 1
        if BitCount == 0            ; Was all the data sent?
            Return Value
            Set INTCON to Just Int

retfie
```

The I/R tank is designed for Sony-brand TV remotes, which have 12 data bits and a 40-KHz carrier. The timings are as listed in Table 9-1 (and use a base timing "T" of 550 µsecs).

Table 9-1. The timings for Sony-brand TV remotes

Feature	"T" timing	Actual length
Leader	4T	2.20 msec
Synch	T	0.55 msec
"0"	T	0.55 msec
"1"	2T	1.10 msec

As I said earlier, one of the features I didn't like about the magazine article robot was its use of a single I/R receiver. I was concerned that this would not provide adequate reception in a crowded room where the robot could be turned in different directions relative to the remote control.

For my robot, I decided to use two I/R receivers pointing 180° apart (this can be seen in the photograph; the I/R receivers are the two square metal cans at the front of the robot). This gives almost 360° coverage.

By doing this, a new problem came up: the problem of arbitration. I found that one receiver might not pick up the transmitted signal or that it would be a cycle or two ahead or behind the other receiver. The arbitration scheme that I used is quite simple; the first receiver to transmit the leader to the PIC will be the one that is "listened" to for the remainder of the data packet.

The final code is shown in Fig. 9-10. This code is "Test9A," which was the last in the series of programs written to understand how TV remote controls work and then build up the understanding into the final application. One thing that I am proud of about this application code is that nothing like an oscilloscope was used to debug it. Once I had enough information to understand how the remote control worked, I was able to write the code.

9-10 The final code for the I/R tank.

```
title   "TEST9A - Control the Robot, at 4 MHz"
;
;  Now, I've finally gotten successful I/R reads for two receivers.
;   Let's see if we can make the robot move.
;
;  This is TEST9 modified to run with a 4-MHz ceramic resonator (not
;   the 1 MHz clock).
;
;  Hardware notes:
;   Reset is tied directly to Vcc and PWRT is enabled.
;   A 4-MHz ceramic resonator is used
;   A 200-ohm resistor and LED are attached to PORTA.4
;   NPN/PNP motor controls ("H" Drivers) are attached to PORTA.3:0
;   An I/R receiver and power hardware is attached to PORTB.7:6
;   A Sony transmitter is used for the data.
;
;  Data stream:
;
;    ____        __  __        __  __      __
;   |    |  :____:  |  |_:  :_____:  |__|  |__...
;
;  State   "0"  "1" "2" "1"  "2"  "1" "2" "1"  - Value of State Variable
;
;  There are 12 bits in the data stream. Values have Been found
;   empirically (program Test5).
;
;
;   Myke Predko
;   96.12.01
;
   LIST P=16C84, R=DEC
   errorlevel 0,-305
   INCLUDE "c:\MPLAB\p16c84.inc"

;  Registers
_w       EQU   0x0C               ;  Interrupt handler save values
_status  EQU   0x0D
State    EQU   0x0E               ;  State of data coming in
Count    EQU   0x0F               ;  Transition counter
Mask     EQU   0x010              ;  Mask byte (only handle one Rx'r)
Last     EQU   0x011
Value    EQU   0x012              ;  Get the two bytes of the value
Valuehi  EQU   0x013
Timer    EQU   0x014              ;  16-bit delay timer
```

```
Timerhi EQU    0x015
   __CONFIG _CP_OFF & _WDT_OFF & _XT_OSC & _PWRTE_ON

   PAGE
;  Code for TEST9A

   org    0
   goto   Mainline

   org    4
   goto   Int
```

; Character read control table

```
STOP      EQU    0x01F              ; Turn OFF the motors and LED

GET_Read                           ; Get controls for the motors
   addwf      PCL                  ; Treat as a table
   retlw      0x0E                 ; 0 - "1" Forward, left
   retlw      0x0A                 ; 1 - "2" Forward
   retlw      0x0B                 ; 2 - "3" Forward, right
   retlw      0x06                 ; 3 - "4" Turn left
   retlw      STOP                 ; 4 - "5" All stop
   retlw      0x09                 ; 5 - "6" Turn right
   retlw      0x06                 ; 6 - "7" Backwards, left
   retlw      0x05                 ; 7 - "8" Backwards
   retlw      0x07                 ; 8 - "9" Backwards, right
```

; Interrupt handler

```
Int

   movwf      _w                   ; Standard interrupt handler code
   swapf      STATUS, w
   movwf      _status

   btfsc      INTCON, RBIF         ; Was the timer engaged?
    goto      Int_Change           ; No, look at the values

   movlw      0x0FF                ; Reset everything
   movwf      Count

   clrf       State

   movlw      8                    ; Reset interrupt (wait for leader)
   movwf      INTCON

   movlw      0x0C0                ; Reset the mask value
   movwf      Mask

   andwf      PORTB, w             ; Save the current value
   movwf      Last

   goto       Int_End              ; Wait for the next change

Int_Change                         ; Something changed, chart it
```

9-10 The final code for the I/R tank. *Continued.*

```
        movf        PORTB, w            ; See if we have a change
        andwf       Mask, w
        xorwf       Last, w
        btfss       STATUS, Z           ; If Zero flag is set, no change
         goto       Int_Chg2            ; No change, clear change bit

        bcf         INTCON, RBIF

        goto        Int_End             ; Wait for the expected to change

Int_Chg2                                ; Now, process the change

        xorwf       Last                ; Get the last value to show

        movf        State, w            ; Handle as a state machine
        addwf       PCL
        goto        Int_Start           ; Setup the leader
        goto        Int_StartHi         ; Leader high value
        goto        Int_Bit             ; Gone low, set up for the bit
        goto        Int_BitHi           ; Read the bit

Int_Start                               ; 0 - Leader, wait for transition

        incf        State               ; Go to the next state level

        movlw       255 - 0x0AA         ; Put in timeout delay
        movwf       TMR0

        movlw       0x028               ; Enable port change AND timer ints
        movwf       INTCON

        clrf        Value               ; Clear out the destination
        clrf        Valuehi

        btfsc       PORTB, 7            ; Setup the mask value
         goto       Int_Bit6            ; Clear bit 6

        bcf         Mask, 6             ; Else, bit 7 set as changing value

        goto        Int_Set             ; Set up the value

Int_Bit6                                ; Bit 8 is set as changing value

        bcf         Mask, 7

Int_Set                                 ; Put in the "last" register value

        movf        PORTB, w
        andwf       Mask, w
        movwf       Last

        goto        Int_End

Int_StartHi                             ; 1 - First high, handle It
```

```
        incf        State               ;   Jump to handling the high value

        movlw       255 - 35            ;   Setup the timer
        movwf       TMR0

        movlw       0x028               ;   Enable interrupts
        movwf       INTCON

        movlw       12                  ;   Set up the counter
        movwf       Count

        goto        Int_End

Int_Bit                                 ;   2 - Low, set up for the timer

        incf        State               ;   Go to ReadCode on next transition

        movlw       255 - 90            ;   Set up the timer
        movwf       TMR0

        movlw       0x028               ;   Enable the interrupts
        movwf       INTCON

        goto        Int_End

Int_BitHi                               ;   3 - Bit value now high, read bit

        decf        State               ;   Want to go down

        bcf         STATUS, C           ;   Shift the value
        rrf         Valuehi
        rrf         Value

        movlw       230                 ;   See if we have a "1"
        subwf       TMR0, w
        btfsc       STATUS, C
         bsf        Valuehi, 3          ;   Set the last bit

        movlw       255 - 35            ;   Reset the timer and interrupts
        movwf       TMR0

        movlw       0x028
        movwf       INTCON

        decfsz      Count               ;   Are we finished?
         goto       Int_End

        clrf        State               ;   Yes, wait for the next value

        movlw       0x0C0               ;   Reset the mask value
        movwf       Mask
        andwf       PORTB, w            ;   Save the current value
        movwf       Last

        movlw       8                   ;   Just wait for commands
        movwf       INTCON
```

9-10 The final code for the I/R tank. *Continued.*

```
Int_End                             ;  Interrupt is complete

  swapf      _status, w             ;  Restore the interrupt
  movwf      STATUS
  swapf      _w
  swapf      _w, w
  retfie

  PAGE
;  Mainline Code

Mainline

  movlw      0x0FF                  ;  Turn off the LEDs
  movwf      PORTB

  movlw      STOP                   ;  Turn off the motor controls
  movwf      PORTA

  bsf        STATUS, RP0
  clrf       TRISA & 0x07F
  movlw      0x0C0                  ;  Just want PORTB.7 as input
  movwf      TRISB & 0x07F
  movlw      0x0D3                  ;  Set up the timer for 16x
  movwf      OPTION_REG & 0x07F     ;   prescaler, not 4x
  bcf        STATUS, RP0

  clrf       Timer                  ;  Clear the 16-bit timer
  clrf       Timerhi

  movlw      0x0FF                  ;  Set up the count
  movwf      Count

  clrf       State                  ;  Start with nothing

  movlw      0x0C0                  ;  Set up the mask and last
                                    ;  value reg
  movwf      Mask
  andwf      PORTB, w
  movwf      Last

  movlw      0x088                  ;  Enable the port change
                                    ;  interrupt
  movwf            INTCON

Loop                                ;  Wait for state = = 2 and then
                                    ;   display data
  movf       Count                  ;  Have we read everything in?
  btfsc      STATUS, Z
   goto      IR_Read                ;  Yes...

  decfsz     Timer                  ;  Wait to turn off motor
                                    ;  controls
```

```
        goto        Loop
        decfsz      Timerhi
        goto        Loop

        movlw       STOP            ;   Turn OFF the motors and LED
        movwf       PORTA

        goto        Loop

IR_Read                             ;   Now, process the value read in

        movf        Value, w        ;   Use the least significant 8 bits

        decf        Count           ;   Have value

        addlw       0x080           ;   Do we have something starting
        btfss       STATUS, C       ;     with 0x080?
        goto        Loop            ;   Nope, wait for the next
                                        character

        addlw       0 - 9           ;   Do we have key "1" - "9"
                                        pressed?

        btfsc       STATUS, C
        goto        Loop

        addlw       9               ;   Restore the value to 0 - 9

        call        GET_Read

        movwf       PORTA           ;   Save the control bits

        movlw       0x061           ;   Set up the 1/2 second wait
        movwf       Timer
        movlw       0x051
        movwf       Timerhi

        goto        Loop            ;   Wait for the next character

        end
```

The controller to be used is a Universal Remote Control set to Sony TV. The I/R receiver the code was written for is a "LiteOn 40-KHz" I/R Remote Control Receiver module (Digi-Key Part Number LT1060-ND). All the other parts are quite standard.

With this code working, I thought I was away to the races.

As a robot platform, I bought two Tamiya "Tracked Vehicle Parts Kit" (Item 70029). Each one has a single electric motor and gearbox; buying two gave me a motor and gearbox for each side's tracks.

Running the motors, I found they used about 250 mA, which was within the stated current ranges of the "H" Drive used in the magazine robot's circuit. I then went off happily building the magazine article's "H" Drive circuit.

Maybe I should explain what an "H" Drive is. It is a balanced circuit that will allow an electric motor to run forward or reverse with only two controls. (See Fig. 9-11.)

When forward or reverse is engaged, a current path will pass through the motor in a specific direction. Looking at the circuit in Fig. 9-11, it should be obvious that *both* forward and reverse should *never* be engaged (else there will be two short circuits between Vcc and Gnd).

After hooking up the recommended circuit, nothing happened except that the transistors got very hot (one actually exploded). This lead to (literally) several weeks of asking questions and experimenting with different circuits (I even pulled out some of my old transistor textbooks; when I do that, I'm *really* desperate) until I determined that I had a circuit that wouldn't work for this application.

To make a long story short, I decided to see what others had done before me. By looking up robotics Web sites (such as the Seattle Robotics Society), I discovered the L293D chip, which is a single integrated circuit that provides the "H" Drive func-

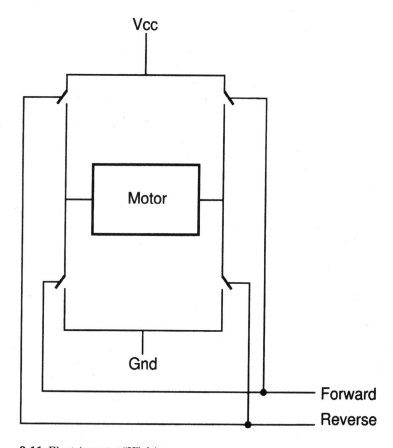

9-11 Electric motor "H" drive.

tion for two motors (making this application much easier to wire than the original circuit, which used eight discrete transistors and resistors).

The L293D consists of four drivers (with clamping diodes) that are meant to be used as "H" Drives for two motors.

Now, with everything working properly, I came up with the circuit shown in Fig. 9-12.

Power is provided by four "AA" alkaline batteries. With Ni-Cads, the robot is very sluggish. Trying to figure out the cause of the sluggish performance, I found that the tracks and axles were binding because they weren't exactly square.

The Tamiya tracked vehicle kit provides a piece of hardwood to put the various pieces on. This is an extremely hard piece of wood and didn't take well to precision placement of parts. I ended up using the piece from the second kit and had better results.

The ultimate end to this was to buy a Tamiya "Wall Hugging Mouse" kit (Item 70068-1300), which contains two motors, two gear boxes, and a plastic chassis to keep everything aligned. In this case, the robot worked a lot better (giving my kids something to terrorize our cats with).

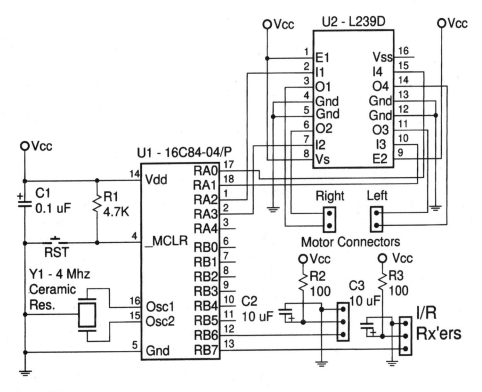

9-12 I/R/ tank schematic.

PROG36: Addendum to the I/R tank

Sometime after I completed the I/R-controlled tank (and the kids were well on their way to demolishing it), a question came up on the PICLIST on alternative methods of reading infra-red remote-control transmissions. The question was about sampling/learning the received data packet rather than comparing it to expected results. The question sparked the idea for representing the codes inside the PIC in other ways.

My first attempt at this was to count the numbers of ones during a sample period. Knowing that a characteristic pattern of ones and zeros would be sampled.

This was the beginning of "IRLCD_2" in PROG36. The program used the schematic shown in Fig. 9-13 on a protoboard.

Note that this circuit uses the serial-LCD interface described earlier in this chapter. Using the LCD really made debugging this experiment (I wouldn't go so far as to call it a "project") a lot easier and really showed me the use of simple interfaces. It's unfortunate that I didn't have my programmer running when I was doing this, because I could have used the programmer's built-in RS-232 port for exactly the same purpose.

The main body of the code, where the I/R stream is read/sampled is shown in Fig. 9-14. This code simply counts the number of ones and stores it in "ReadCount" for a given amount of time. The theory behind this method of sampling was that the dead space between packets would be read along with the data and the result would combine them.

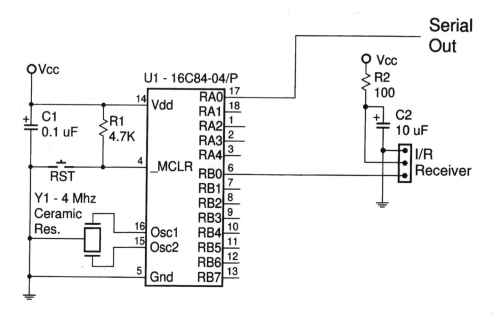

9-13 I/R test circuit.

9-14 The main body where the I/R stream is read/sampled.

```
        movlw      0x0A0           ; Setup the timer interrupt
        movwf      INTCON

Loop                              ; Loop here for each update of the
                                  ; screen

        movlw      200            ; Wait for the time out
        subwf      IntCount, w
        btfss      STATUS, Z
        goto       Loop           ; Has NOT timed out

        movlw      200            ; Can we display?
        subwf      ReadCount, w
        btfsc      STATUS, Z
        goto       Loop_Reset     ; Reset the count values

        movf       ReadCount, w   ; Now, display what was read in

        clrf       IntCount       ; Clear the display values
        clrf       ReadCount

        call       DispHex        ; Display the hex value

        movlw      0x08E          ; Reset the cursor for writing
        call       WriteINS

        goto       Loop           ; Wait for the next loop around

Loop_Reset

        clrf       IntCount       ; Reset the values
        clrf       ReadCount

        goto       Loop


Int                               ; Interrupt, check I/R input

        movwf      _w             ; Save the context registers
        swapf      STATUS, w
        movwf      _status

        bcf        INTCON, T0IF   ; Clear the timer interrupt

        incf       IntCount       ; Increment the count register

        btfsc      PORTB, 6       ; Increment the read value?
        incf       ReadCount

        movlw      256 - 25       ; Reset the timer
        movwf      TMR0

        swapf      _status, w     ; Restore the context registers
```

9-14 The main body where the I/R stream is read/sampled. *Continued*

```
movwf       STATUS
swapf       _w
swapf       _w, w
retfie
```

The actual value returned from the program wasn't very repeatable (as was expected). For example, five tries with the "1" key from a universal remote programmed with Sony codes produced these results:

0x09F
0x09D
0x08C
0x09D
0x09D

Generally, the results from this program were repeatable about 60% of the time. This might have been acceptable except for the poor discrimination that this method had. For example, the codes for "2" and "3" are 0x081 and 0x082, respectively. The problem lies in the fact that the two codes have the same number of ones and zeros. The code *might* pick up the differences, but I didn't find this to be the case.

So, the code for reading the I/R packet was changed to what is shown in Fig. 9-15.

9-15 The modified code for reading the I/R packet.

```
        clrf        IntCount        ; Reset the counters
        clrf        ReadCount

GetPack                             ; Get the next packet coming in

        movlw       0x088           ; Wait for port change interrupt
        movwf       INTCON

Loop                               ; Loop here for each update of the
                                   ; screen

        movlw       150             ; Wait for 25 msec of data from
                                   ; I/R
        subwf       IntCount, w
        btfss       STATUS, Z
        goto        Loop            ; Has NOT timed out

        clrf        INTCON          ; No more interrupts for a while

        movf        ReadCount, w    ; Get the read in CRC

        clrf        IntCount        ; Reset for the next packet
        clrf        ReadCount

        call        DispHex         ; Now, display the character
```

```
        movlw      0x08E              ;  Reset the cursor
        call       WriteINS

        goto       GetPack            ;  Wait for the next I/R packet

Int                                   ;  Interrupt, check I/R input

        movwf      _w                 ;  Save the context registers
        swapf      STATUS, w
        movwf      _status

        movlw      0x020              ;  Just wait for a timer interrupt
        movwf      INTCON

        movlw      256 - 20           ;  Reset the timer
        movwf      TMR0

        incf       IntCount           ;  Increment the count register

        bcf        STATUS, C          ;  Now, figure out what to add to LSB
        btfsc      PORTB, 6           ;  Is the incoming value set?
        goto       Int_Set

        btfsc      ReadCount, 5       ;  Do we update the value coming in?
        bsf        TATUS, C

        goto       Int_End

Int_Set                               ;  Incoming set
        btfss      ReadCount, 5       ;  Is the current bit set?
        bsf        STATUS, C          ;  No, turn on the incoming bit

Int_End

        rlf        ReadCount          ;  Shift over with new input data

        swapf      _status, w         ;  Restore the context registers
        movwf      STATUS
        swapf      _w
        swapf      _w, w
        retfie
```

The fundamental changes were: The sampling started after the Leader was received and the 1s and 0s were treated as the inputs to a Linear Feedback Shift Register. In the Christmas projects earlier in this chapter, I explained briefly how Linear Feedback Shift Registers work. For the code in Fig. 9-15, an 8-bit LFSR was used to produce CRC codes. In this case, the input wasn't the high bit of the shift register, instead it is the input from the I/R receiver.

Using this code, the CRC codes listed in Table 9-2 were generated from the Sony I/R transmitter.

Table 9-2.
The CRC codes

Key	Code
Power	0x052
Vol+	0x05E
Vol−	0x0BB
Ch+	0x0DC
Ch−	0x062
"0"	0x017
"1"	0x07A
"2"	0x08D
"3"	0x033
"4"	0x01F
"5"	0x04E
"6"	0x072
"7"	0x0CC
"8"	0x0B9
"9"	0x023

The interrupt handler code waits for a Port Change interrupt (the I/R line going low from its nominal state of 1). Once that happens, the line is sampled every 200 µsec, and a CRC is generated from each sample. After 150 samples (30 msec), the CRC is output serially in hex format (i.e., sending the high nybble followed by the low one).

The CRC generated is rock solid (none of the 60% repeatability I had with just sampling bits). I don't know if I'm going to go back and update my I/R tank code (lack of initiative more than anything else), but this is clearly a *much* more elegant and robust method of handling I/R codes.

I did a limited amount of checking for invalid code rejection by reprogramming my universal remote with Panasonic and RCA codes. The CRC's generated were different from the Sony ones given in Table 9-2.

I was never pleased with the XORing used to create the CRC in the previous code. I felt that it was too confusing to understand. After some thought, I came up with the idea that, if I used the same bit number for the PORT input bit as the CRC "tap," I could simplify the CRC generator (from "bcf STATUS, C" to "rlf ReadCount" in Fig. 9-15) to:

```
bcf        STATUS, C
movf       PORTB, w         ; Get the value read in
xorwf      ReadCount, w     ; XOR it with the current
andlw      0x040            ; Clear all the bits but the two
btfss      STATUS, Z        ; we're interested and if not = 0
 bsf       STATUS, C        ; then make the LSB of the CRC = 1
rlf        ReadCount
```

This code only improves the original by two addresses, but it sure is a lot easier to understand!

When I did this, I did cheat a bit and changed the CRC tap (and not the line coming in, which would have meant I would have to change the LCD code). However, it still ran very well, with unique CRCs generated for each of the different keys of the Sony mimicking universal remote.

I guess the moral of this whole escapade is that tremendous improvements in your code (in terms of size and effort requirements) can be made if you look at a problem from a different direction. The code literally took less than six hours to develop and debug (compared to over two weeks for the I/R receiver of the tank). It takes up about a third of the space of the tank and only uses two 8-bit variables compared to the seven of the Tank's code. This is a tremendous improvement!

Discussing this philosophically, it can be seen that this experiment is actually restructuring the application (reading an I/R transmitter) to best fit the PIC. The data read is now totally 8-bit, as opposed to the 12/16 bits that had to be handled in the original application.

PROG32: Electronic thermometer with 7-segment LED displays

This section outlines how to build a 9-V powered thermometer using a standard PIC and peripheral parts that can be bought (literally) from Radio Shack. (See Fig. 9-16.) No precision components are used in the circuit, which leads to the use of a calibration value that is stored in EEPROM.

9-16 The electronic thermometer.

The thermometer itself consists of a thermistor and three 7-segment LED common cathode displays. The code displays the current temperature in degrees Celsius. The thermistor was bought from Radio Shack (Part Number 271-110).

The circuit shown in Fig. 9-17 was used for the PIC thermometer. Note that the circuit has been built both on a protoboard and a perfboard using point-to-point wiring. No precision components are used in the circuit.

In the schematic, notice the 9-V battery circuit in the top-left corner. This is the circuit used for all the projects that use a 9-V battery (or attach to a "wall wart" through a 9-V battery connector).

For my design (and code), I used the following segment pattern:

```
  —6—
 ¦     ¦
 5     1
 ¦     ¦
  —7—
 ¦     ¦
 4     2
 ¦     ¦
  —3—
```

Each LED segment is connected to a PIC pin via a 220-Ω resistor, except for the segment connected to RA0, in which the 220-Ω resistor is attached to Vdd and the

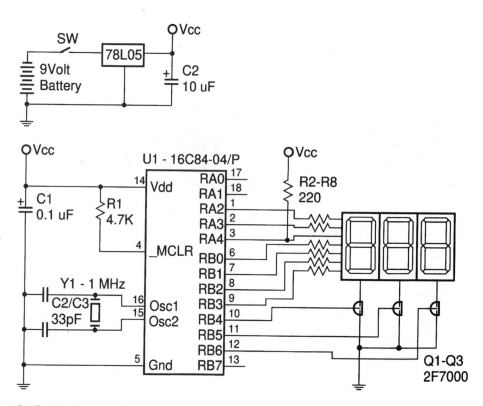

9-17 LED thermometer.

PIC pin pulls it low to turn *off* the LED (RA4 is an open-collector Output, not a totem-pole).

In my application, I wired the segments as:

RA2 is Connected to Segment 6 RA3 is Connected to Segment 5 RA4 is Connected to Segment 1—Note the comments above RB0 is Connected to Segment 7 RB1 is Connected to Segment 4 RB2 is Connected to Segment 3 RB3 is Connected to Segment 2

The reason for using these values was strictly to make the wiring easier. If you look up at the photograph at the top of this section, you'll see that the seven current-limiting resistors are placed between the PIC and the LED displays.

I generally recommend using a protoboard to try out new hardware. While this is a good in general, this circuit is so simple that you might as well just go directly to the final project. I used a small box and matched circuit board with point-to-point wiring. The circuit really only took me a few hours to build.

When doing something like this, you really have to plan ahead. This means that you should be putting down the various components before soldering to make sure everything fits. Even then, I moved around a couple of components to make the wiring a bit easier.

Another trick to making sure that the building will be successful is to develop small, focused test programs to test each facet of the design and assembly. For the digital thermometer, I wrote "Test1" (in subdirectory Prog32). This program simply cycles through the individual segments of the display to make sure they are wired correctly. The program itself uses tables for each of the segments. This means that, to set up both the segments in PORTA and PORTB, only two table reads are required. Test1 does *not* use interrupts for handling the interrupt display.

Test1 is also used for the "wake up" test for the PIC in the application. This allows you to determine whether the voltage regulator is working along with the clock and reset of the PIC.

Next, Test6 can be run; this will allow you to start checking the function of the thermistor/capacitor measuring circuit. The program uses the formula that was discussed elsewhere in the book:

$$time = R \times C \times -\ln\left(\frac{Vend}{Vstart}\right)$$

to measure the resistance of the thermistor. To get the actual temperature, I used idealized components and had the temperature looked up from a table (because the thermistor resistance versus temperature is a nonlinear function, and I didn't want to have to use an extensive mathematical formula). The table relates the resistance value read (where the time taken for the cap to discharge is proportional to the resistance) to an actual temperature. The table values relate back to idealized components (i.e., exact values).

Test6 is also significantly different from Test1 in how it displays values on the LEDs. Test1 was unable to display anything on more than just one display at a time; however, Test6 uses an interrupt handler to toggle between the displays (at 2 KHz) to give the appearance that they are all working at the same time. The interrupt handler makes the multiplexing between displays to be transparent to the mainline code; all it has to do is load up the values to be displayed for each digit, and the interrupt handler takes care of the rest. The thermistor is wired like the pot in chapter 8 (the series resistor is 100 ohms and the capacitor is 0.1 µF).

Once the program starts to run, you will see that it displays a value for the current temperature, or an error message "uuu" or "^^^" for too cold and too hot, respectively. While the circuit is up and running, you might want to fool around with it a bit, like putting the thermistor between your fingers and watching the temperature going up, or putting it into a refrigerator/freezer and watching it go down. In doing this, you will discover that the thermistor-based thermometer is a lot faster than a mercury-based one. That is because the thermistor has a lot smaller thermal mass than the mercury thermometer and can reach the surrounding temperature faster.

As you look at the displayed temperature, you will probably notice two things. The first is that the temperature is probably wrong. That is because you are using components which are *not* perfect (ideal); their values are somewhat off the exact specified values. The second is that you will probably see the temperature creep up if the thermistor is close to the PIC. This temperature creep is caused by the PIC warming up as it operates. Both of these problems can be overcome by calibrating the PIC and the circuit.

One obvious way of calibrating the circuit is to put in a trimmable part with the resistor or capacitor. The downside to this method is that the trim parts can drift themselves, either by material breakdown over time or by the trim parts being knocked around.

To avoid these problems, I used the PIC's EEPROM to store in the calibration value. This means that the previous equation was changed to:

$$time = k \times R \times C \times -\ln\left(\frac{Vend}{Vstart}\right)$$

with the calibration value k being a constant.

Test7 uses two pulled up I/O pins, which when pulled down allows the user (you) to enter in different values and have them stored into the PICs EEPROM. These connectors, when shorted will pull the PIC pins that they are associated with to ground (changing the input level from a high to a low). I put in a third of a second delay to allow for debouncing and to give a delay between updating the value. Using MPLAB, you can read back the EEPROM values to see how they change.

A significant feature of Test7 is how the data is stored in the EEPROM. On power up, two data-check bytes are checked for the value 0x0AA and 0x055. These data-check bytes are used to indicate that the value in the data EEPROM byte is actually correct. On the first power up of the PIC with this program, the checksum bytes are set up, and the data EEPROM byte is set to zero. This way, you are starting at a known location. Using the check bytes along with value byte means that, if you shut off the power to the PIC, when you come back later, the value will still be there and usable.

Test7 is also where I encountered a really significant problem in debugging the code. Every time I would run it, only one digit would be displayed—only the first digit of the EEPROM data check was written. After much and protracted debugging, I found the problem was with the initialization of the segment variables. I had copied one line from the previous and not changed its value. This ended up costing me about three weeks of part-time debugging to find the problem. The lesson being that

I didn't make sure my variable initialization was correct before I looked for other problems.

Once Test6 and Test7 were working, I combined them to get PROG32. PROG32 contains the temperature reading and displaying code from Test6 along with the EEPROM read and write code from Test7 to provide a calibration value that can be used to correct the delay value to get an accurate temperature.

The calibration value itself is 16-bits long and is multiplied by the actual delay value. The high byte of the resulting 16-bit number is used as the corrected delay value. Using the high byte is the same as dividing the result by 256 (which is what the calibration value is based on). Doing the calculation this way eliminates the requirement to provide a division routine or floating point routines as part of the calibration.

This method can be used in a number of other applications where calibration or fractions of a value have to be found.

For example, let's say you wanted to find 30% of an 8-bit number. This could be done two different ways. The first would be to multiply by 3 and divide by 10. If we wanted to get 30% of 123 this would be:

$$30\% \text{ of } 123 = \frac{(123 * 3)}{10}$$
$$= \frac{369}{10}$$
$$= 36 \text{ or } 37 \text{ (depending on rounding)}$$

This can be done in the PIC as multiplying 123 by 30% of 256 and just taking the high byte of the 16-bit product like so:

$$30\% \text{ of } 123 = \frac{(123 * 30\% \text{ of } 256)}{0x0100}$$
$$= \frac{(123 * 77)}{0x0100}$$
$$= \frac{9,471}{0x0100}$$
$$= 36 \text{ or } 37 \text{ (depending on rounding)}$$

So, multiplying by fractions is actually a lot easier in PICs (and other digital processors) than you probably first thought.

The 7-segment LED driver code and hardware specified within this application can be used for a variety of purposes. Different character sets (i.e., hex codes) and additional and different displays can be used easily by modifying the code.

PROG33: Marya's Toy—addendum to the LED thermometer

After I built the LED thermometer, my daughter, who was 18 months old at the time, found it absolutely fascinating with its LED displays. In fact, I had a lot of problems trying to keep her from wanting to play with it.

9-18 Marya's Toy.

The solution was to come up with a toy of her own that had lights and buttons that would respond to her inputs. (See Fig. 9-18.)

This was also a good chance for me to experiment with other types of LED displays. The display that I used is a 15-segment alphanumeric display. This display is very similar to the 7-segment displays of the electronic thermometer:

```
   ----1----
 ¦\     ¦    /¦
 6   91011    2
 ¦    \¦/    ¦
   -7-  -8-
 ¦    /¦\    ¦
 5  121314   3
 ¦/    ¦    \¦
   ----4----    Dot(15)
```

except, it has a lot more segments. In this diagram, I show how I numbered each of the segments.

Part of the project was to be able to drive the segments using a 13 I/O PIC (preferably a 16C84).

Actually, this was quite easy to do because the experience I got from doing the "Frosty the Snowman" came into play here. I loaded two shift registers up with the data for each segment, during a TMR0 initiated interrupt handler, and shift data out to display the character.

For selecting which display is active, I used a 74S138, rather than a single transistor. This way, up to eight displays could be handled without additional components. This also made the display selection very simple.

For this project, because I used more displays (6 as opposed to 3 in the thermometer) and the complexity of loading the bits for each display increased, I increased the PIC clock speed to 4 MHz. Also, when the display shift registers are being loaded, I disable the display output, which really eliminates any possibility of flashing displays.

"Prog1" in the "Prog33" subdirectory is the application program that I did. You can see that I have created a scrolling display interface; this was done by having a foreground timer (which is completely independent of the display TMR0 interrupt handler) that sets up output values for each display, and after a set time, it moves the data over.

This project is really an amalgamation of a lot of the other projects that I have done here. For this reason, like the Christmas decorations, I'm not going to go into as full detail as some of the other projects in this chapter.

The only real new thing in this project is the use of the complex LED displays; the design could be used these and other LED displays (e.g., dot-matrix LED displays could be done with the shift register method or selecting different columns like each display is selected in this or the Thermometer).

PROG38: The MIC

The parallax Stamp was a marvelous idea. The original purpose was to create a simple, standalone computer in a very small space. The Stamp was actually how I first got into the PICs; I was fascinated by the device that was able to run the BASIC code.

As great as the Stamp is, I felt there were three shortcomings:

- The need for an external chip to store the program in.
- The need for a PC to "compile" the code before loading it into the Stamp.
- The relatively slow execution speed (about 1000 to 2000 instructions per second)

As I researched the PIC microcontrollers, I discovered the 16C84 with its built-in EEPROM memory and wondered if I could create my own Stamp and avoid the problems I saw.

Rather than jumping right into developing my own BASIC language that would fit inside the 1K address space, I decided to see what I could do with a "simulated/emulated" microprocessor. For the most part, I feel that I was successful in creating a project without these shortcomings.

The project presented here is not complete; I found that I ran out of space before I could implement the final feature (back checking for labels when entering Assembler instructions). However, the project shown here will show you what can be done with a single chip microcontroller.

I should also point out that this project served as the basis for the programmer and emulator presented elsewhere in this book. It was really this project that gave me the confidence to realize that the PIC is capable of much more than just being a programmable piece of logic; it can also be used to interface intelligently with the person using it.

Like all projects, the first order of business was to decide what I would call the finished product. "MIC" jumped right out at me.

With this out of the way, I then began to try to decide what the emulated device's architecture would be. I don't know if the way I did it was putting the cart before the horse (I've never designed a computer architecture before), but I based the design on what I wanted the instructions to look like when they where displayed for the programmer.

I should point out that the idea was for a single chip microcontroller that used a Harvard architecture similar to the PIC; which is to say it provides its own program storage (using the 16C84's 64 bytes of EEPROM data memory), along with file registers and an 8-bit I/O port. Along with this, I wanted the chip to have its own assembler and disassembler with a simple monitor to allow single stepping and program execution. Obviously, this is a lot to ask for in only 1024 instructions.

The design I came up with is a true 8-bit processor: All data paths are 8-bits wide, and there are a maximum of only 8 address bits. (See Fig. 9-19.)

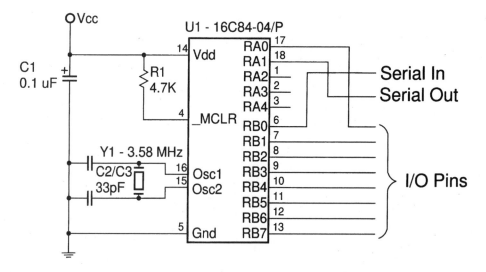

9-19 "MIC" circuit.

The prompt that I came up with for the user is:

```
!Label PC ACC Flags Ins > _
```

where *!Label* is the label at the current Address, *PC* is the Program Counter, *ACC* is the accumulator, *Flags* is the Zero and Carry flags, and *Ins* is the disassembled instruction at the Program Counter.

You will probably note the similarity of this prompt to the one I used for the Emu.

To make the instructions more intuitive, I wanted everything to be based upon the accumulator. For example, the instruction:

```
+ 77
```

would be the addition operation:

```
A = A + 0x077
```

Table 9-3. The MIC instruction set

Instruction	Description
<	Load the accumulator
>	Store accumulator in register
+	Add to contents of accumulator (store result in accumulator)
−	Subtract from contents of accumulator (result in accumulator)
¦	OR with contents of accumulator (store result in accumulator)
&	AND with contents of accumulator (store result in accumulator)
^	XOR with contents of accumulator (store result in accumulator)
/	Shift register to the right by 1 (store result in accumulator)
#	Skip the next instruction if the accumulator bit is set
@	Jump to the specified address, store address + 1 in "B"

With this format, I decided upon the 10 instructions listed in Table 9-3 for the MIC. These instructions gave me all the necessary functions I could think of (with some warnings).

A shift left of a register would be accomplished by adding a value to itself:

```
< Reg                    ;  Load accumulator with Reg
+ Reg                    ;  Reg + Reg = Reg << 1
```

Branching on condition would be carried out by moving the flags register contents into the accumulator ("A" register) and then skipping on the bit. A skip on carry set would be:

```
< F                      ;  Load the accumulator with the flags register
# 0                      ;  Skip if Carry bit set
```

You'll also notice there isn't a "call" instruction. Instead any time a goto ("@") instruction is executed, the incremented Program Counter is stored in "B." This makes the return from a subroutine:

```
< B                      ;  Get the return value
> C                      ;  Change the Program Counter
```

Doing subroutines this way saved code and file registers needed for a program counter stack.

With the instruction set specified, I then decided on how to do the register addressing.

I decided upon three modes: immediate, register address, and register indirect. To show how each would be used adding a value to the accumulator:

```
+ 37                     ;  Add 0x037 to contents of accumulator
+ D                      ;  Add contents of the "D" register to accumulator
+ [D]                    ;  Add contents of register addressed by "D"
                         ;   to the accumulator
```

If you look back at the instructions, you'll see that each instruction takes data in these formats and acts on the value they represent.

In specifying the registers, I decided upon eight "base" registers and three special-purpose registers. The base registers (A to H) all can be written to and read from, but there are some special purposes assigned to some of the registers (Table 9-4).

Table 9-4. The special purposes of the registers

Register	Special purpose
A	Accumulator (the destination for all arithmetic operations)
B	"Goto" return address
C	Program Counter
F	Flags register

The Program Counter "C" can only contain values between 1 and 0x01F. If a value outside these limits is written to this register, program execution will stop. (This was put in to provide a software-initiated stop to the program.)

There are a maximum of 31 instructions in the MIC. This is because I use two bytes to store each instruction in the 16C84's EEPROM. The first byte contains two check bits (to tell the MIC that the EEPROM is programmed) along with two bits for the addressing mode and four bits for the instruction. The second byte of the instruction is the value that the instruction uses.

The "F" or flags register contains the result of arithmetic STATUS along with the serial receive/transmit status.

The Flags bits are arranged as shown in Table 9-5. The high for bits is the complement of their low nybble counterpart to allow the skip on bit set to work for all conditions.

Table 9-5. The arrangement of the Flag bits

Bit	Function
0	Carry flag, set on addition > 0x0100 or subtraction < 0
1	Zero flag, set on arithmetic result = 0
2	Receiver byte waiting
3	Transmitter free
4	Not Carry flag
5	Not Zero flag
6	Not receiver byte waiting
7	Not transmitter free

The three special-purpose registers are the "P" and "T" registers (which are really the PORT and TRIS registers of the PIC) and the "X" register. The "X" file (sorry, I couldn't resist) register, when written to, transmits the byte and, when read from, reads the current character (or 0x000 if there isn't an unread character). As the code is currently written, the "Transmitter Free" bit will never be reset because the transmission code happens in line, which means that program execution won't resume until the character transmission has completed.

Once this definition was completed, I was able to create the block diagram shown in Fig. 9-20.

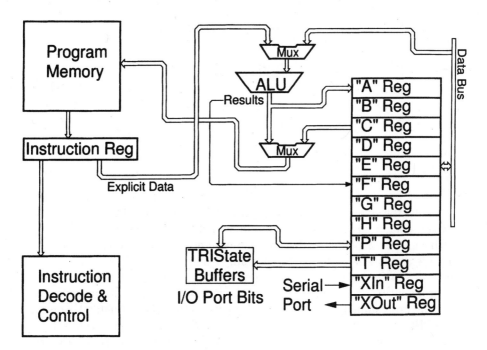

9-20 Full "MIC" architecture.

Now, that the instructions had been defined and I had a processor architecture, I went ahead and began to write the code.

If you go through the various source programs, you will see how I developed first the serial interface and line parser, next loading the program into EEPROM, followed by accessing the registers, and finally the execution routines. Everything in this project works except for the back annotation of the labels.

For example, if you were entering the code:

```
@ !B                     ;  Jump on to label "!B"
 .
 :
!B
```

when the "!B" label was entered, the code doesn't go back looking for unassigned references to it.

The operational instructions I came up with for the monitor are:

```
Reg                  ;  Display the contents of the register
Reg = Value          ;  Change the contents of the register
Ins                  ;  Put the instruction at the current "C" value
!A - !E              ;  Set the label
1 - Blank Line       ;  Single Step
R                    ;  Run the program from the current "C" value
```

Here is a simple test program that you might want to try (it outputs an incrementing number and puts in a delay before doing the next):

```
< 0                  ;  Enable all the outputs
> T
> P                  ;  Save the current output value
```

```
!A                          ; Return here for each loop
  < 0                       ; Wait for a counter to equal 0 for the delay
  > D
!B                          ; Delay loop
  < D                       ; Increment the counter
  + 1
  > D
  < F                       ; Check for zero
  # 1
  @ !B                      ; Not zero, loop again
  < P                       ; Increment the output value
  + 1
  > P
  @ !A                      ; Go to the start of the loop and start again
```

Actually, this is quite an intriguing project from the idea that you have a completely self-contained development system on a chip that costs less than $10. Looking at this and then looking at the programmer and emulator, you'll see where a lot of the code and interfacing ideas came from.

If I were to do this project over again, what would I change to improve on it? I would probably change the serial receive/transmit to use an algorithm similar to that used by the programmer (i.e., sample at three times the data frequency). This would free up RB0 for use with the other 7 bits of PORTB and would simplify the code somewhat. In doing this, I would probably streamline the data receive to use the line buffer as the received data buffer (rather than contending with two separate data buffers in which data is transferred from one to the other).

The last change I would make to this project is to add a serial EEPROM for external program storage (get a full 255 instructions available to the MIC; I would still save 0x000 or 0x0FF as a stop address for the software).

PROG40: Keyboard interface

The typical example shown for interfacing a microcontroller to a switch-matrix keyboard is a 4×4 keypad. While this can be useful in some applications, often only a full "qwerty" keyboard will do. For this project, I wanted to go through how a switch-matrix keyboard could be attached to a PIC. (See Fig. 9-21.)

A switch-matrix keyboard is a series of switches wired in rows and columns. They can be read by pulling up each row individually, tying down a column to ground, then seeing if a switch is pulling row to ground through the column. (See Fig. 9-22.)

The pull-down transistors shown in Fig. 9-22 can be either discrete transistors or PIC I/O pins individually enabled to a zero state to simulate the pull down to ground.

From this graphic, it can be seen that when one of the switches is closed and the column it's connected to is pulled down to ground, the receiver will sense a low, where if the switch was open, the it would sense a high value. This method of switch sensing is analogous to having multiple open-collector outputs on a single pulled-up line.

9-21 The keyboard interface.

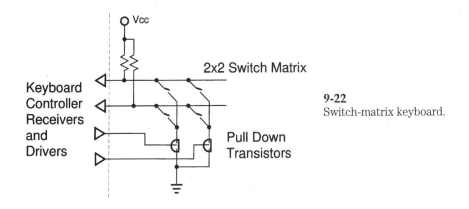

9-22
Switch-matrix keyboard.

To "scan" the keyboard, each pull-down is individually enabled in series, and any time an input value is low, the key at that row/column address is deemed to have been pressed.

With this scheme, there are a few issues to consider:

- How should the keyboard be wired to the controller?
- What about switch bouncing?
- What about multiple keys pressed at the same time?
- What about Shift/Ctrl/Alt/Function key modifiers?

In developing the program "Key1," I had to understand these issues and have a plan to deal with them.

Trying to figure out how a keyboard should be wired when you don't have any information is not as daunting a task as it would appear to be. When I bought the keyboard used for this project, I got it from a surplus store for $1. The only identifying feature on it was a strip indicating pin 1 on a 15-pin connector.

The first thing that I did was set up a matrix, and using a DMM, I "beeped" out every key with the two different connector pins. With this information, I created the matrix shown in Fig. 9-23. Once this was done, I manipulated the table until I could get a good understanding of how the keyboard was wired and what would be the best way to attach it to a PIC.

The design point I decided upon was setting up eight rows (or register bits) for each column. I defined the "row" as where I put the pull up and the "column" as the pin I pull to ground.

The data was transformed into the table shown in Fig. 9-24, where the rows and columns are the pin numbers on the connector.

With this information, I was ready to specify the wiring. At this point, it was simply wiring the rows and columns to the PIC (with the "rows" having the pull up as noted previously).

I used a 16C57 as the PIC because it had more than enough I/O pins to handle the 15 pins required by the keyboard. I outputted the data (through a MAX232) to

9-23 The keyboard matrix.

Pin	1	2	3	4	5	6	7	8	9	10	11	12	13	14	15
1					"V"	"R"			"4"		"M"	"J"	"F"	"7"	"U"
2					"C"	"E"			"3"		","	"k"	"D"	"8"	"I"
3					"X"	"W"			"2"		"."	"L"	"S"	"9"	"O"
4						Ctrl			Fctn		"="	" "	"Shft		Ent
5											"B"	"Z"			
6							"T"	"Q"							
7									"5"		"N"	"H"	"G"	"6"	"Y"
8									"1"		"/"	";"	"A"	"0"	"P"
9													Caps		
10															
11															
12															
13															
14															
15															

9-24 The data in tabular form.

Row/Column	5	6	9	11	12	13	14	15
1	"V"	"R"	"4"	"M"	"J"	"F"	"7"	"U"
2	"C"	"E"	"3"	","	"K"	"D"	"8"	"I"
3	"X"	"W"	"2"	"."	"L"	"S"	"9"	"O"
4		Ctrl	Func	"="	" "	Shift		Enter
7	"B"	"T"	"5"	"N"	"H"	"G"	"6"	"Y"
8		"Q"	"1"	"/"	";"	"A"	"0"	"P"
10			Caps					

my PC's RS-232 terminal emulator program so that I could monitor what was coming out (and, if needed, send debug information as well).

To read the keys, I used the algorithm shown in Fig. 9-25.

9-25 The algorithm to read the keys.

```
KeyPreviously = 0                       ;  Nothing currently read

Loop

   Dlay4ms                              ;  Delay for key debouncing

   if KeyPreviously = 0
      KeyCount = 0                      ;  Reset # of times through loop
      For i = 0 to # columns and No KeyPreviously
         Scan Column                    ;  Check each column and set
                                        ;   KeyPreviously
   else                                 ;  else key previously Pressed
      if KeyPreviously Still Pressed
         KeyCount = KeyCount + 1         ;  Increment the actual count
         if KeyCount = = 5               ;  Do we have the first press?
            Send KeyPreviously
         else
            if KeyCount = 128            ;  Have we waited 1 sec?
               Send KeyPreviously
            else
               if KeyCount = 192         ;  Have we waited another 1/2 sec?
                  Send KeyPreviously
                  KeyCount = 128         ;  Reset for next autosend
      else                              ;  Key was lifted
         KeyPreviously = 0              ;   Start all over

   goto      Loop
```

This algorithm handles the key bouncing by requiring that 5 consecutive polls 4 msec apart "see" the key being pressed. When the bit goes high, the key read is reset until the next keypress pulls down a row.

The "Scan Column" routine resolves multiple keypresses (which can obviously happen when more than two keys in the same column are pressed at the same time) by taking the lowest active bit in the column.

The "Send" routine looks up the ASCII code to send by reading the value in a row/column table.

It should be noted that the key modifiers (Ctrl, Func, etc.) are always masked off in "Key1." I didn't bother to put in the modifiers because I didn't have an application that required the keyboard (and being lazy...). They can be implemented easily by adding new tables for each modifier and then, before calling the table to look up the value, adding an offset to the correct table values.

Note that I wouldn't bother debouncing the key modifiers because they are not the action that initiates the action of sending the keys. They are just used to make sure the proper ASCII codes are sent.

The code presented here ("Key1.asm") could very easily be ported to a mid-range PIC with the advantage that the TMR0 interrupt could be used to initiate the scan, allowing the code to be run totally in the background.

PROG41: Servo controller

The servo controller (Fig. 9-26) is an excellent example of what can be done with the PIC. The servo controller uses a 16C71 to provide an user interface with an LCD and allows the user to control up to four servos, develop a sequence of events for the servos to run, and allow the user to save a sequence for later execution.

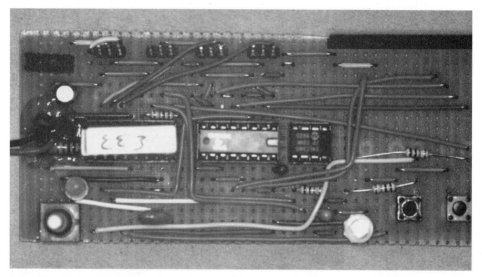

9-26 The servo controller.

Pretty good for an 18-pin device.

This project is designed for controlling the servos used in "armature" robots or mechanical displays that require moving parts. The servos can either be controlled individually or sequenced. (See Fig. 9-27.)

I apologize for the cluttered schematic, but I wanted to make sure that all the devices the PIC is interfacing to can be shown in the graphic. One thing to notice is that a lot of the lines are actually Vcc or Ground. If all the power lines were taken off the schematic, it would be a lot simpler.

Actually, when I first developed the circuit (and the schematic), I just went ahead and assembled it on a vectorboard (as shown in Fig. 9-26). A vectorboard is a good prototyping tool for this project because of the repeated Power and Grounds, which are easily bussed using the vectorboard.

The user interface consists of a Pot, two buttons, and a 16-character by two-line LCD. Making it all work together is a menu routine that takes a source message, puts it on the screen, then waits for and handles user input.

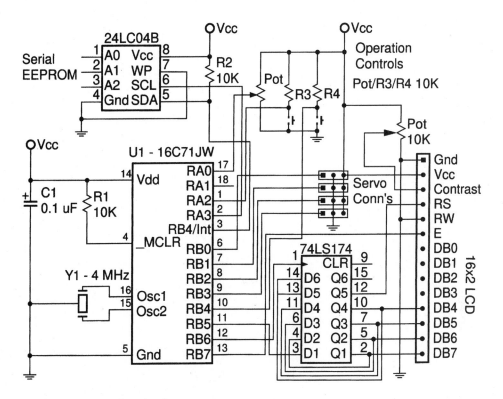

9-27 Servo controller circuit.

The initial screen looks like:

```
Servo Controller
>Pos< Pgm  Run
```

The button at RB4 is the "Select" button, which will move the cursor (the ">" and "<") between the actions on the lower line of the LCD. When the "Enter" button (at RA2) is pressed, the display program is ended and the cursor position is returned to the caller.

The Pot is used to select arbitrary values (i.e., the servo position). This was the first time that I worked with the 16C7x analog inputs (the Pot was used as a voltage divider).

The code for reading the Pot value is really quite simple:

```
ADCON1 = 2                    ;  Just RA0/RA1 for analog I/P
ADCON = 0x041                 ;  Enable A/D conversions for RA0
.
:
ADCON.GO = 1                  ;  Start the A/D conversion
Loop While ADCON.GO = = 1     ;  Wait for the A/D to complete

ADTemp = ADRES                ;  Read the A/D result
```

This code must seem almost unbelievably simple considering how complex it seems in the datasheets, but it really does work. The conversion takes about 20 µsecs to complete.

To be totally fair, the A/D conversion is simple because I am measuring a single, (mostly) dc voltage. This code would be a bit more complex if ac signals (with frequencies in the audio range) were measured. However, it could still be quite easily done.

Using a Pot to specify exact values is kind of tricky and will require some practice and patience. An optimal solution would be using a multi-turn Pot instead of the single-turn one that I used (you should also put on a knob to make turning the pot easier on the fingers). The reason why I went with the Pot in the first place is because the servo position and program delay values are not really precision operations. The addition of the "goto" and "print" functions to the program really were an afterthought.

The user interface is used to provide a nonlanguage-based programming environment. With the Pot and "Select" and "Enter" buttons, you can specify an immediate servo position, enter in a program, or run the program (either single step or running with 20-msec nominal steps). I believe that it's a very efficient interface for applications such as this.

The servo interface is actually embarrassingly simple. I used radio-control model servos for this project, which rely on a 1-msec to 2-msec pulse (the duration specifies the position) every 20 msec. The 20-msec cycle is a natural for a TMR0-based interrupt handler. When invoked, the interrupt handler outputs a pulse of 1 msec to each servo and then loops with a counter. When the counter value is greater than the value for a particular servo, the pulse for that servo is turned off. The counter continues to loop until a full 2 msec (to allow full travel of a servo) is complete. TMR0 is then reset to an 18-msec delay (so the whole cycle is 20 msec long).

The mainline program is responsible for updating the servo positions.

Note that, for this project, the servo position granularity is such that there are 50 steps from stop to stop. This is due to the parallel control of the servos. The count loop takes quite a long period of time checking each servo value to see if it has to be updated. The number of steps can be increased (the granularity decreased) by either using a faster PIC clock or by using fewer servos and taking out the code used to support the unneeded devices.

The last major subblock of this project is the I²C serial EEPROM. The 24LC04B contains 4 Kbits, or 512 bytes. For each program instruction, I use two bytes, the first six bits of the first byte is a check value (because instructions that have an invalid check will stop the running program), followed by two bits for the instruction type (Set Servo Position, Goto Location, Delay n/10 seconds, and Print Character on the LCD). The second byte contains the value to be used by the instruction.

When the program is running, it can be stopped manually by pressing the "Enter" button.

The PIC itself communicates with the serial EEPROM by behaving as an I²C master. The 24LC04B is an I²C slave device, which means it responds to instructions directed to it by a master (these instructions are prefaced by a control byte).

Rather than give a full explanation of how I²C works with regard to serial EEPROMs, I just want to discuss some aspects of I²C as it relates to this project.

The two most important things to note are that, with the 24LC04B, the ninth bit of the address is in the control byte (normally address bits over the eighth one are contained in a separate data packet (as is done with the EEPROM used in the "Emu"). The other thing to note is that the A0 to A2 pins on the EEPROM package are *not* connected to the chip inside. This means the 24LC04B cannot be used with other EEPROM devices (which typically use the A0 to A2 bits to differentiate each other when the control byte is sent) on the I²C bus because of the danger of contention (two devices each trying to transmit data).

The actual code itself for communicating with the EEPROM is quite simple and, as it's laid out in "ServoEE3.asm," makes reading and writing to the EEPROM simply just subroutine calls.

By adding together the LCD/Pot and button interface to the servo controller and serial EEPROM, you really end up with a project that is more than the sum of its parts. I think you'll find the interface intuitive and easy to work with. This was my two boys first experience in programming (they are 8 and 10), and they were able to master the four-instruction programming environment very easily.

One of the things that made the project a lot more accessible to children was mounting the servos on "Lego" blocks (if you're a household with kids, chances are you have a few hundred pounds of the stuff). (See Fig. 9-28.)

9-28 Lego/servo combination.

PROG41: Addendum—SimmStick servo controller

There are a lot of PIC development tools out there (my programmer is one) for the PIC. One thing I did want to show you was how some of these tools can really speed up your application development.

The "SimmStick" (Fig. 9-29) by SiStudios is one such product that provides a platform for the PIC with oscillator, power supply, reset, serial EEPROM, and RS-232 interface and even leaves space for a prototyping area. The SimmStick is also Basic Stamp compatible, so you can see there's a lot on it!

To show how easy it is to set up one of these for a project, I decided to redo one of my more complex projects (the servo controller) on a SimmStick.

The whole operation went quite easily with me only having to cut two traces going from RA0 and RA1 and adding wires from RA3 and RA4 to make the SimmStick compatible with the servo controller application. (If I wasn't using the built-in A/D on the 16C71, I probably wouldn't have even done this; I just would have changed the code to using whatever pins the SimmStick has already specified for this purpose with traces.)

Once the traces were cut and wires were put in to change the EEPROM use, I went ahead and built the SimmStick for the servo controller application. I stopped and made sure that the PIC would run (basically check Osc2 to make sure it was running).

Using wire wrap techniques, I then put on the 74LS174 that's used as a shift register for the LCD and the two buttons and a 10K pot. Once this was done, I then created a simple vectorboard wiring of the application.

Total time to create the application was about 5 hours. This includes the time required to read through the documentation and decide how I would wire everything. For example, rather than have wires running across the back of the SimmStick and making it hard to follow the added wiring, I put the buttons and Pot out to an unused pin on the

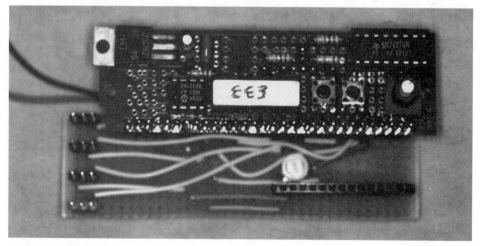

9-29 SimmStick servo controller.

SimmStick and then ran the signals through the vectorboard. If I was to do another application using the SimmStick, the time required would be reduced significantly.

If you compare the SimmStick version of the servo controller to the original vectorboard version. You'll see that the wiring is quite a bit simpler and the whole package is reduced significantly in size. I highly recommend products like the SimmStick for developing applications because of their ability to simplify wiring (basically eliminating the PIC set up—like the oscillator and power supply wiring from being an issue).

10
CHAPTER

Emulators

Emulators tend to be quite costly and can be difficult for the beginner to use. They can, however, provide significant advantages in debugging an application. Unlike a simulator, an emulator is capable of running actual hardware at full functional speed. It will also allow you to play "what if" with the hardware, without having to go through the trouble of reprogramming a PIC. Emulators are invaluable tools for getting applications out on a deadline, but they are generally not required for the hobbyist and are certainly not required for any of the applications in this book.

The PICMaster

I was very lucky to be lent a PICMaster for my use during the writing of this book. While I didn't use it for debugging any of the applications, I did get to spend enough time with the PICMaster to understand what it is capable of. To summarize my feelings, I would have to say that the PICMaster is the Cadillac of PIC ICE's.

This is primarily for two reasons:

- *The integration of the PICMaster with the MPLAB development environment.* The PICMaster works exactly like the MPLAB simulator (but with more features), which minimizes the learning required to use it.
- *The PICMaster was designed for (and by engineers).* As I worked with the unit, I didn't find anything that I would change (and no wishes for more function).

For example, say you were working on a serial communications application and you have the situation when, after an instruction is sent out, the wrong character *seems* to have been sent out. Because of the single failing instruction is inside a number of other instructions, you have trouble isolating the problem.

Using the PICMaster, you could set a trigger point for the instruction received, which would then trigger a digitizing scope, record the PIC instructions that were executed, and then stop when the instruction response had completed.

In this example, once the event has completed, you have an instruction flow record along with a scope picture of the actual electrical signals (and this can be re-run repeatedly with varied parameters to look at different aspects of the problem).

This, by the way is not a fictitious example; I've done many things like this before on a variety of processors using their respective ICEs.

The heart of the PICMaster is a Bondout PIC chip. The Bondout chip is an actual PIC chip (or *die*), less a package, which has extra connections between the Bondout pads and the PICMaster hardware. These Bondout pads allow the PICMaster hardware direct access to the data, address, and control busses. This means the PICMaster can not only monitor what is happening in the program, but also control execution and register contents.

As I said previously, the PICMaster can be used as a "what if" tool. If you were working on a project and wanted to try something new or take a different approach to a problem; this would be the tool to used.

Having said this, I should now put in the warning that Iteel the PICMaster (or any ICE) is not a development tool. An ICE won't fix problems that are caused by poor design or incorrect algorithms. In fact, before powering up the emulator and trying your code out on hardware, you should have simulated your code in every way possible to eliminate any possible defects.

The PICMaster is a debug tool; it will help find the causes in an application.

A few points to note about the PICMaster. The device itself consists of a PC ISA card connected to the PICMaster, which is connected to a probe (which is plugged into the PIC socket of the application). The ISA card has an address switch that must be set to a valid, unused address range (as explained in the manual) before it can run properly.

I found the pod to be quite difficult to work with (not seeming to make good contact in the PIC socket); although to be fair, this could be because of the hardware I was using. At the time of this writing, the PICMaster only works in Windows 3.1 or Windows '95 environments. However, all told, the PICMaster would be a part of my "dream" PIC development system.

PROG42: The Emu

After lusting over the PICMaster, I decided to try to design my own simple emulator. Not having access to Bondout chips and not wanting to develop a logic solution of my own (i.e., designing a PIC in VHDL and programming it into an FPGA), I decided to come up with my own simple design. (See Figs. 10-1 and 10-2.)

This design takes the .hex file from a program (downloaded from a terminal emulator on a host PC in a manner that is similarly to the programmer). The program is loaded into a I^2C serial EEPROM, taking up 2 bytes of storage for each instruction.

The emulator presented here will emulate the 16C61 and 16C71 (and 16C84 without the EEPROM data) and could do low-end 18-pin PICs (if the source was re-assembled for the 16C61 and a reset "goto" was put at address 0).

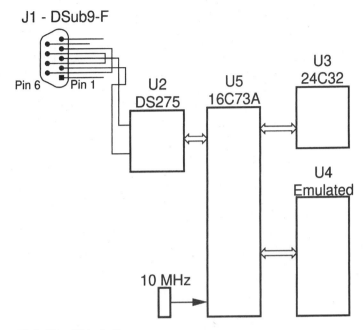

10-1 "Emu" block diagram.

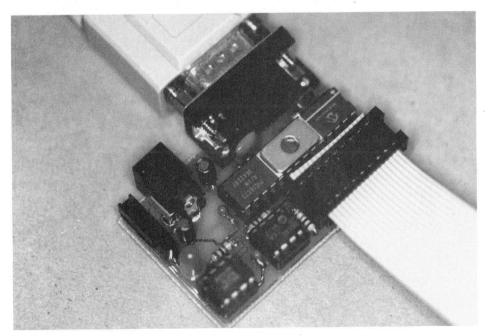

10-2 The Emu.

What's truly amazing about this project is that it really doesn't use any hardware or interfaces that haven't been discussed earlier in the book other than the use of the 16C73A's SCI port for serial communication.

Actually, the largest problem I had writing this program was balancing the mainline and subroutines between the two program store pages. When you look at the code (EmuTest7.asm), you'll see that, at the end of the first page and beginning of the second page, I've placed all the far call code (no jumps between the pages). This code can be shown in the general case:

```
movlw    HIGH FarRtn
movwf    PCLATH
call     FarRnt
movlw    HIGH $
movwf    PCLATH
```

This meant that I could not pass data back and forth using the "w" register. This could be improved by manipulating the pclath bits directly using "bcf"/"bsf".

The Emu itself uses the first five PORTA bits and all the PORTB bits to emulate the I/O bits of the PIC. Except for one ADCON (16C7*x* A/D) bit mismatch, which is compensated for in software, the hardware registers are pretty much indentical to the emulated device's.

The TMR0, OPTION, INTCON, and PCL/PCLATH registers all required some software massaging.

The TMR0 register is either incremented by each instruction step or by an external pulse on the RB0/Int line. To emulate this, a separate 16-bit counter was set up (the 16-bits to handle the prescaler) to be used along with the 16C73A's TMR0 (which was used was specified by the "T0CS" bit in the OPTION register).

Actually, I didn't emulate the OPTION register or its function, but it is accessed a lot to make sure that the TMR0 is running properly.

INTCON is totally emulated. The reason for this was that I didn't want the emulated PIC to cause an interrupt in the 16C73A, which would make the program execution more complex by having to handled unexpected interrupts and discriminate between them and expected TMR1 (the RTC interrupt) and the SCI data received interrupt. The E_INTCON IF flags are checked before an instruction is run, and then the emulated interrupt request is made.

The emulated PCL and PCLATH and corresponding Program Counter registers work exactly like their real counterparts:

- In "gotos" and "calls," the address in the instruction is loaded directly in to the PC, and PCL is updated.
- For PCL updates (e.g., "addwf PCL"), the value in PCLATH is put into the high value of the Program Counter.

Not surprisingly, the emulated instructions are very easy to code:

```
addwf    Reg, w
```

is done by:

```
movlw    0x0F8        ;  Clear the status bits
andwf    E_STATUS
call     ReadReg      ;  Read the specified register
```

```
addwf    E_w, w      ;  Execute the instruction
movwf    ResTemp     ;  Store the result temporarily
movf     STATUS, w   ;  Update the status bits
andlw    7
call     WriteReg    ;  Store the result in the correct register
```

When the program is running, the Emu first fetches the instruction from the EEPROM and then executes it. During execution, the EEPROM is left in Sequential Read Mode to avoid having to send the overhead of a control byte, a new address, and a read command.

A "goto"/"call" requires a new address to read the instruction at the new address. The extra overhead of doing the control byte, new address, and read almost perfectly scales the length of time for the "goto"/ "call" as twice that of a regular instruction.

Because of the need to get the instruction from the EEPROM, the effective speed of the emulator is about 10 KHz (the EEPROM is accessed at 200 KHz). The software was originally designed for a 20-MHz 16C73A clock, but the 20-MHz ceramic resonators I had on hand had too much internal capacitance for the clock to run.

The user interface is very similar to that of the MIC. Unlike the MIC, it is capable of communicating with the host at 9600 bps.

If you remember the problems with the programmer data speed, you might be surprised to find out that the Emu has no issues with 9600 bps data transfers. This is because the 24C32 (the EEPROM used) has a "Block" write command that allows up to 64 bytes to be stored before initiating the 10-msec write cycle. As noted in the text of my programmer, the maximum number of instructions that are stored on a line of the .hex file is 16, so I can load up a block of memory for programming and then initiate the write cycle while waiting for the next line to be received.

Another feature that I take advantage of with the EEPROM is the fact that, when the initial control byte is sent, the acknowledge bit (pulled low when the EEPROM is ready for the next instruction) is polled before any read/write operation. This eliminates the need to write a delay routine for the worst case write. Instead, the 16C73A polls the EEPROM until it is able to accept the next command.

Another advantage of the Emu over the programmer is that there is no 50 instruction jump limit. This is because the address to be written to is sent to the EEPROM before the data (eliminating the PC increment instructions of the programmer).

Operating the Emu is pretty simple, the operating commands are quite straight forward. (See Table 10-1.)

Like the programmer, I have provided raw card information for the Emu (Fig. 10-3) with a component placement drawing (Fig. 10-4).

Unlike the programmer, there are plans for making it available through Wirz Electronics (the address is listed in appendix D) and licensed dealers. While the design presented here works, I suspect that the released version will have additional features (20-MHz 16C73A speed, more PIC devices, full interrupt support, etc.).

This is not to say that the version presented here doesn't work. The "servo controller" project was used to debug the Emu, and a number of minor problems were found.

Table 10-1. The Emu operating commands

Instruction	Description
D 61¦71	PIC device
1 [*address*]	Single step
S [*address*]	Step over call
X [*address*]	Execute, stop on keystroke
P *address*	Set PC
E *register*	Display/change register
B [*address*]	Toggle BP
U [*address*]	Unassemble
A [*address*]	Assemble
R	Display regs
T [*value*]	Display/set timer
W [1¦0]	Display/enable WDT
C [R¦L¦X¦H]	Clock used
L	Load EEPROM
!	Reset the emulated device

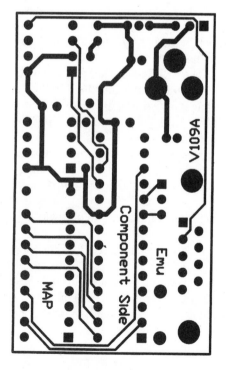
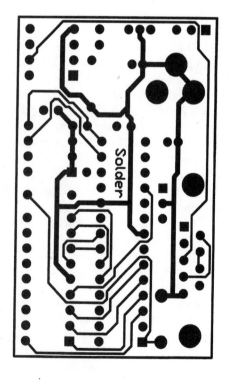

10-3 Emulator raw card redesign.

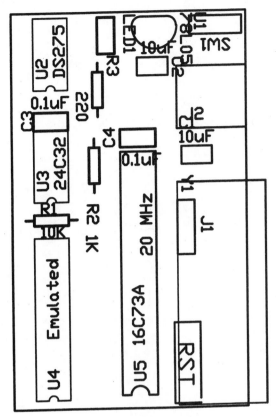

10-4 Emulator component layout.

A

Hints

One thing I hate about most technical books is that there's never a list of the important items that are mentioned in passing.

In this book, I wanted to list out the 10 most important hints that I have found for writing software, executing PIC programs, and interfacing to other devices. I hope this will help enable your first (and subsequent) applications work correctly right from the start and help you avoid having to learn these simple hints the hard way.

If you do have problems, here are some lists of things to check for.

Software hints

1. Initialize all RAM Registers. Upon power-up, the RAM registers can be any value. Avoid problems with them by making sure they are initialized. The first thing I do in all my software is to initialize the RAM registers that I am going to use.
2. When using tables or state machines, make sure that interrupts are disabled (or you are in the interrupt handler, which amounts to the same thing) before executing the "addwf PCL,f" instruction. This is to prevent an interrupt from pushing the incorrect next address on the PC stack.
3. After a "sleep" instruction which you are planning to wake up from, put a "nop" following the "sleep." This will ensure that no unexpected or invalid instruction is executed upon wake up.
4. If you are using "incfsz" or "decfsz," make sure the instructions are used on registers that can equal zero. Some registers in some processors have bits which are always set to one.
5. Subtraction of registers, in the PIC, can be confusing. Subtraction of a constant can be downright unreasonable. To make subtracting a constant from a register easier to understand, Add the *negative* of the constant (i.e., 0 − constant) instead.

6. In the PIC 16C5*x* family, make sure no subroutine labels are located in the upper 256 addresses of a 512 address page. This problem can be eliminated if the subroutine label is put in the lower 256 addresses and a "goto" subroutine is put in following it.

7. Make sure that the watchdog timer is explicitly turned off in the "__CONFIG" statement and when a PIC is programmed. If the watchdog timer is inadvertently enabled during programming of the device, then time-outs can cause the PIC to seemingly reset periodically.

8. Check the maximum depth of the Program Counter (the maximum number of nested subroutines both inside the mainline and the interrupt handler), and make sure that the program does not exceed the depth of the PIC's Program Counter stack. If the stack is exceeded, then the program will never return to the highest caller.

9. Do not use absolute addresses instead of labels or relative addresses. This will make it very difficult to reuse code later.

10. The instructions "movf w, w" or "movf w, f" will assemble correctly, but will not execute as expected in the low-end and mid-range PICs. In the PIC include files, "w" is defined as 0 (zero), and this value will be put in as the register address. This means that in actuality, you are executing the instruction "movf INDF, w|f".

PIC execution hints

1. If a simple reset circuit is to be used, put in a current limiting resistor (5K to 10K) between Vcc and _MCLR to allow easy reset of the PIC by a momentarily on push button (or a handy screwdriver) shorting _MCLR to an adjacent Ground pin.

2. If the PIC just sits there and does nothing when power is applied, first check to see that the PIC clock is running (check OSC2) using logic or oscilloscope probe.

3. If the PIC seems to reset itself periodically during execution, check for the proper use of the decoupling cap and verify that the watchdog timer is not timing out.

4. Don't set the code-protect bit in the configuration register unless you're absolutely sure. Use of this feature could prevent reuse of the device at a later time.

5. When programming the EEPROM data registers in the 16F84, use the example code, and make sure the registers are accessed as shown in the datasheets, because the timing is critical to proper execution of the EEPROM.

6. When programming the EEPROM of the 16C84, make sure that the EEDATA register is set before the EEADR register. Once this is completed, the data can be programmed into the EEPROM.

7. If you have an application that has critical timing requirements, which are controlled by TMR0, it is recommended that you *never* write to this

register. Doing so clears the prescaler (if it's feeding TMR0), and you might end up with unpredictable results in terms of time intervals.

8. When developing an application, put in the hardware required to allow in-circuit programming of the device. This can give you a significant board manufacturing cost reduction.

9. In an interrupt handler, if you are waiting for a timer to overflow, check the Interrupt Active Flags before executing the timer overflow handler routine. If TMR0 is free running, then the T0IF bit in INTCON will be set if the timer overflows. (Unlike other Interrupt Active flags, it is not only enabled when the interrupt-enable bit is set.)

10. Always put a Ground pin in your design to allow you to have a constant ground for scoping and hooking up a logic probe.

Interfacing hints

1. Initialize port output bits before changing the TRIS register. This will prevent bit values from being at an unexpected state after the write to the TRIS register.

2. Most instructions (including "bcf," "bsf," "clrf," and "movwf,") which would seem to not have to read the port actually do read the port before putting a value into it.

3. If you are using a input line with multiple sources, make sure the line is pulled up to Vcc, all the switches strictly pull the line to ground and all logic outputs are open collector. If this is not done, then there is an opportunity for bus contention and an invalid logic level being driven to the input pin.

4. When you are interfacing to another device's RS-232 port and you just want to have a three wire connection (Rx, Tx, and Gnd), the modem control bits can be tied together in the manner shown in Table A-1 to fool the host's RS-232 serial port into seeing the handshaking signals that it expects.

Table A-1. How the modem control bits can be tied together

Lines	25-pin D Shell	9-pin D-Shell
RTS - CTS	4-5	7-8
DTR - DTS	6-20	4-6

5. An open collector output can be simulated by using a regular (tri-statable) output programmed with an output value of 0. To put the bit in the open collector "high" state (not pulling down the signal), the bit is in Input Mode. By enabling a pin to output with 0 already in the PORT register's corresponding bit, the net is pulled down to a "low" state.

6. When driving any low-impedance device (e.g., LEDs), always make sure a current limiting resistor (100 Ω to 200 Ω) is used. The PIC can only source/sink a set amount of current (usually around 20 mAmps); however, to be on

the safe side, make sure the PIC isn't exceeding the current limit for the output device.

7. PICs should sink (pull down to ground) a signal rather than source (drive to Vcc) the signal. This is because the PIC is able to sink more current than it can source.

8. When using a PORTB pin for reading a potentiometer value (i.e., emulating the Basic Stamp "Pot" command), do not enable the internal PIC pull-ups because they will change the linear operation of the RC network.

9. The PIC clock can be fanned out to other devices in the circuit. However, only the signal from OSC2 should be used.

10. Transmitting RS-232 data to another device should only be done using an RS-232 level translator (e.g., Maxim MAX232). This will ensure that all possible devices will be able to receive the signal.

16-bit operations

As you probably noticed in chapters 8 and 9, I often find 8-bit numbers insufficient for the purpose at hand. 16-bit (and larger) numbers can be handled easily in the PIC, even though it is an 8-bit processor (only able to move 8 bits at a time).

So that you don't have to read through the text to try to find every incident of 16-bit data handling, I've tried to list them here for you along with a few extras. These snippits of code could be made into macros or put into libraries to make your coding easier. While I haven't listed every possible operation, the various routines can be built upon to create whatever functions are required.

Defining 16-bit numbers

I define 16-bit numbers in a manner similar to that of an 8-bit number and just give them 2 bytes in the RAM register space. The following example shows how to define an 8-bit variable, followed by two 16-bit variables:

```
RAM          equ     12          ;  Start of RAM for the PIC16C71
Reg_8        equ     RAM         ;  Define the 8-bit register
Reg_16       equ     RAM + 1     ;  Define the first 16-bit register
Reg2_16      equ     RAM + 3     ;  Define the second 16-bit register
```

or, using the "CBLOCK" command:

```
CBLOCK 12                        ;  Start of RAM for the 16C71
Reg_8                            ;  Define the 8-bit register
Reg_16, Reg_16HI                 ;  Define the first 16-bit register
Reg2_16, Reg2_16HI               ;  Define the second 16-bit register
ENDC
```

Note that "Reg2_16" is two addresses above the start of "Reg_16." This is to give "Reg_16" 2 bytes. To maintain consistency, I access the high byte of the 16-bit variable by using the name and adding one to it. For example, to access the high byte of "Reg_16," I use "Reg_16 + 1".

For the "CBLOCK" example, note that I have defined each high byte with the "HI" label. This eliminates the need for accessing the high byte as:

```
Reg_16 + 1
```

In these examples, I alternate between the two methods of defining 16-bit variables.

Note that I haven't included operations on stack variables. It is not very difficult to carry them out, but the resulting operations are dependant on the implementation used and how the FSR register is set up and handled.

The algorithms and code in this appendix are all in 16 bit two's complement format. This means that you can traverse between positive and negative numbers and allow them to interact with each other easily. It should be noted that, if a positive number becomes greater than 32K −1, it will become negative as far as these routines are concerned. Care must be taken to make sure the range limits are not exceeded; otherwise, the values will change.

Increments and decrements

Incrementing a 16-bit Value is very simple:

```
incf        Reg              ;  Increment the low byte
btfsc       STATUS, Z        ;  Do we have zero (multiple of 256)?
  incf      Reg + 1          ;  Increment high byte (if necessary)
```

The decrement of a 16-bit value isn't quite so simple:

```
movlw       1                ;  Will be subtracting "1" from variable
subwf       Reg
btfss       STATUS, C        ;  If Carry flag is NOT set, decrement
  decf      RegHI            ;    the high byte
```

Because the "decf" operation only sets the Zero flag upon completion, there is no way of knowing whether or not a borrow of the high byte is required after decrementing the low byte. By using the "subwf" instruction, the Carry flag is set when a decrement is not required (allowing the skip over the high byte decrement).

Addition/subtraction of constants

Addition and subtraction of 16-bit variables to constants can be done in a very similar manner. There is one trick, however, and this is to always do the high byte first.

The reason for doing the operation on the high byte first is to ensure that the destination has the final result. If the low byte requires a carry, the high byte can be incremented/decremented easily, without having to add extra instructions to do the high byte.

So for adding a constant to a value:

```
Reg = Reg + 0x01234
```

the following code is used:

```
movlw       0x012            ;  Add the high byte first
addwf       Reg + 1
movlw       0x034            ;  Add the low byte next
addwf       Reg
btfsc       STATUS, C        ;  Don't inc high byte if carry reset
  incf      Reg + 1
```

This code can be cut down according to whether or not the high or low byte of the constant is equal to zero.

The corresponding subtraction:

```
Reg = Reg - 0x01234
```

looks like:

```
movlw        0x012           ;  Subtract the high byte first
subwf        RegHI
movlw        0x034           ;  Subtract the low byte next
subwf        Reg
btfss        STATUS, C       ;  Don't dec high byte if carry set
 decf        RegHI
```

If you are adding and subtracting to a 16-bit variable and storing the result in another variable, then basically the same code can be used. The only difference is that, rather than specifying the "Source" destination in the addwf/subwf instructions, the "w" register is specified and the result is stored in the destination:

```
Destination = Source + 0x05678
```

will look like:

```
movlw        0x056               ;  Add high byte first
addwf        Source + 1, w
movwf        DestinationHI       ;  Store result in destination
movlw        0x078               ;  Add low byte next
addwf        Source + 1, w
movwf        Destination         ;  Store result
btfsc        STATUS, C           ;  Is the Carry Flag set?
 incf        DestinationHI       ;   Yes, increment high byte
```

Subtraction with the result being stored somewhere else is carried out exactly the same way.

Addition/subtraction of other variables

Addition of a 16-bit variable to another 16-bit variable is similar to that of adding a constant to a 16-bit variable. If the destination is the same as one of the values:

```
a = a + b
```

the code looks like:

```
movf         bHI, w              ;  Add the high bytes
addwf        a + 1
movf         b, w                ;  Add the low bytes
addwf        a
btfsc        STATUS, C           ;  Put in the carry
 incf        a + 1
```

If the destination is different from both values to be added:

```
c = a + b
```

the code is changed to save the sums in "w" and then store it in "c" like:

```
movf         a + 1, w            ;  Add the high bytes
addwf        bHI, w
movwf        cHI
movf         a, w                ;  Add the low bytes
addwf        b, w
```

```
movwf       c
btfsc       STATUS, C              ;  Increment due to carry
 incf       cHI
```

Subtraction is carried out in the same way, but care must be taken to ensure that the subtracting register is kept straight (something that is less of an issue with addition) and put in "w" before the subtraction. If you want to do the following statement:

```
c = a - b
```

you would use the code:

```
movf        bHI, w                ;  Get value to be subtracted
subwf       a + 1, w              ;  Do the high byte
movwf       cHI
movf        b, w                  ;  Get the value to be subbed
subwf       a, w
movwf       c
btfss       STATUS, C             ;  Look for the carry
 decf       cHI
```

Other operations on constants and variables

Doing other operations (bitwise or whatever) on 16-bit values can use the code shown earlier as a base. The big difference between the code in this section and the earlier code is that you don't have to worry about the carry flag.

For example, ANDing a 16-bit variable with 0x0A5A5 would be done like:

```
movlw       0x0A5                 ;  Get value for ANDING
andwf       Reg + 1               ;  Do the high byte
andwf       Reg                   ;  Do the low byte
```

This follows on for the other types of operations (i.e., with another 16-bit variable or with a different destination).

There is one difference, however, and that is to do with rotating 16-bit values. Rotating must be carried out in such a way that the Carry flag is always correct for the shift.

This means that the Carry flag must first be cleared (to put a 0 in the bit getting the Carry flag), then the first rotate should be selected in such a way that the second will have a valid Carry flag (i.e., holding the value to be transferred from the first register).

For example, to shift left:

```
bcf         STATUS, C             ;  Clear the Carry flag for new bit
rlf         Reg                   ;  Shift the low byte
rlf         Reg + 1               ;  Shift high byte with carry from first
```

To shift right:

```
bcf         STATUS, C             ;  Clear the Carry flag for the new bit
rrf         RegHI                 ;  Shift down the high byte
rrf         Reg                   ;  Shift down low byte with valid carry
```

Comparisons with 16-bit variables

Comparisons involving 16-bit variables require that the comparison value (register) is subtracted from the register to be checked. The results of this will then tell you what is going on with the condition. I use the same code as is shown earlier and save the result in temporary values and then look at the result. The subtraction code used for comparing a 16-bit variable to another 16-bit variable is:

```
movf      Reg2HI, w          ; Get the high byte of the result
subwf     Reg1HI, w
movwf     _2                 ; Store in a temporary register
movf      Reg2, w            ; Get the low byte
subwf     Reg1, w
btfss     STATUS, C          ; Decrement high if necessary
 decf     _2
```

At the end of this series of instructions, "w" contains Reg2-Reg1 and "_2" contains Reg2HI-Reg1HI with the borrow result of Reg2-Reg1.

If the variable is to be compared against an immediate value, then the "movf" instructions would be replaced with "movlw" and the two bytes of the immediate value.

There are six basic conditions that you can look for: equal, not equal, greater than, greater than or equal, less than, and less than or equal. So, to discover whether or not I have any of these conditions, I add the following code.

For equal and not equal, the value in "w" is ORed with "_2" to see if the result is equal to zero:

```
iorwf     _2, w              ; Is the result = = 0?
```

For equal add the lines:

```
btfss     STATUS, Z          ; Execute following code if == 0
 goto     Zero_Skip          ; Else, code != 0, skip over
```

For not equal, append:

```
btfsc     STATUS, Z          ; Execute following if != 0
 goto     NotZero_Skip       ; Else, code = = 0, skip over
```

If a greater than (the 16-bit variable is greater than the comparison value), then the result will not be less than zero (two's complement negative). Actually, the same code (just with a different bit skip) can be used to test for greater than:

```
btfsc     _2, 7              ; Not negative, 16-bit is greater
 goto     NotGreater_Skip    ; Else, skip if not greater Than
iorwf     _2, w              ; Is it equal to zero?
btfsc     STATUS, z          ; No, it is greater than
 goto     NotGreater_Skip    ; Else, if zero, not greater than
```

Note that just the most significant bit of the 16-bit difference is checked. If this bit is set (equal to 1), then the 16-bit variable is less than the comparison. If it is reset (equal to 0), then it is greater than and you should check to see if the result is not equal to zero (or else it is equal).

For less than:

```
btfss     _2, 7              ; Negative, 16-bit is less than
 goto     NotLess_Skip       ; Else, skip because it's not less than
```

To check for greater or equal to, the last three lines of the code checking for greater than are simply erased. To check for less or equal to, the three lines from not equal to are added before the check for less than.

Here is the complete code for compare and skip on Reg1 less than or equal to Reg2:

```
movf      Reg2HI, w          ; Get the high byte of the result
subwf     Reg1HI, w
```

```
movwf        _2                    ; Store in a temporary register
movf         Reg2, w               ; Get the low byte
subwf        Reg1, w
btfss        STATUS, C             ; Decrement high if necessary
  decf       _2
iorwf        _2, w                 ; Check for equal to zero
btfsc        STATUS, Z             ; If not zero, jump over
  goto       EqualLess_Skip        ; Equals, jump to the code
btfsc        _2, 7                 ; If number is negative, execute code
  goto       EqualLess_Skip        ; Else, jump over
```

Multiplication

16-bit multiplication uses a bit-shift algorithm to create a product from two values. Each bit of the multiplier is checked to see if it is set, and if it is, a corresponding multiplicand is added to the product.

In pseudo-code the algorithm is shown in Fig. B-1. This pseudo-code is actually designed to be easily translatable to PIC instructions. The algorithm in Fig. B-2 works for both positive and negative numbers.

B-1 The multiplication algorithm in pseudo-code.

```
Product = 0
LoopCount = 16                          - Do for each bit

Loop                                    - Come back here for each
                                          bit
  if ( multiplier & 1 ) != 0            - Can add the bit
    Product = Product + multiplicand

  multiplier = multiplier >> 1          - Shift down the multiplier
  multiplicand = multiplicand << 1      - Shift up multiplicand for
                                          next add

  LoopCount = LoopCount - 1
  if LoopCount != 0                     - Do 16x
    goto Loop
```

B-2 The multiplication algorithm in Assembler.

```
clrf         Product               ; Initialize the product
clrf         Product + 1

movlw        16                    ; Initialize the loop counter
movwf        LoopCount

Loop                               ; Come back here for each bit
  rrf        Multiplier + 1        ; Put the LSB into the carry register
  rrf        Multiplier            ; And shift over the multiplier
  btfss      STATUS, C             ; If Carry set, add the multiplicand
    goto     AddSkip               ; Else, skip over the add
```

```
    movf    Multiplicand + 1, w   ;  Add multiplicand to product
    addwf   Product + 1
    movf    Multiplicand, w
    addwf   Product
    btfsc   STATUS, C
    incf    Product + 1

AddSkip                           ;  Now, shift over multiplicand/loop back
    bcf     STATUS, C
    rlf     Multiplicand
    rlf     Multiplicand + 1

    decfsz  LoopCount             ;  Have we done it 16x?
    goto    Loop                  ;  No, keep looping
```

At the end of this code, Product will have the value of Multiplicand time Multiplier. Note that three 16-bit variables are required to execute this algorithm.

This routine can be cut down or scaled up according to the numbers to be used.

Division

I have not included a general division routine because there isn't a general algorithm that handles both positive and negative numbers and passes back both the quotient and remainder efficiently. In my experience, I have found that implementing a division routine is very dependent on the expected values. There are some very efficient algorithms for specific divisors, the most obvious being how to divide by multiples of two (simply shift the value to the right).

Along with the problems with negative values and returning the required value, handling zero can cause problems. In many processors, division by zero causes a system fault.

A general form for a division routine could be the division of the core of the pseudo-code is a bit-shift analogous algorithm to multiplication (Fig. B-3).

B-3 The division algorithm in pseudo-code.

```
if Dividend < 0                   - Change the dividend to a positive
                                    number
  Dividend = 0 - Dividend
  dividendneg = 1                 - Mark that we have to change it
                                    back
else
  dividendneg = 0
if Divisor < 0                    - Repeat with the divisor
  Divisor = 0 - Divisor
  divisorneg = 1
else
  divisorneg = 0
```

B-3 The division algorithm in pseudo-code. *Continued*

```
Count = 0                            -  Going to count where division
                                        starts
Quotient = 0                         -  Store the quotient
while ( Divisor & 0x0400 ) != 0  -  Find the start of the division
  Count = Count + 1                  -  Increment the number of bits
                                        shifted

  Divisor = Divisor << 1

while Count != 0                     -  Now, do the division
  if Dividend >= Divisor       -  A subtract can take place
    Quotient = Quotient + 2 ^ Count
    Dividend = Dividend - Divisor
  Count = Count - 1
  Divisor = Divisor >> 1

if Dividendneg == 1                  -  Now, change the values
  if Divisorneg == 1
    Quotient = Quotient
    Remainder = 0 - Dividend
  else
    Quotient = 0 - Quotient
    Remainder = 0 - Dividend
else                                 -  The dividend was positive
  if Divisorneg == 1
    Quotient = 0 - Quotient
    Remainder = Dividend
  else
    Quotient = Quotient
    Remainder = Dividend
```

Looking through the code in Fig. B-3, you can see that most of the space is taken up with making sure the dividend and quotient polarities are represented correctly.

The core of the algorithm consists of finding the number of bits the divisor has to be shifted to being able to start subtracting efficiently from the dividend and then just going through and dividing the value.

The core can be implemented in a PIC as shown in Fig. B-4. Note that the 16-bit variable "Count" and the 8-bit variable "Temp" are required and that the divisor cannot be equal to zero.

B-4 The division algorithm in Assembler.

```
clrf      Quotient        ;  Initialize the variables
clrf      QuotientHI

movlw     1               ;  Set the count (bit tracking) to 1
movwf     Count
clrf      CountHI

StartLoop                 ;  Find the starting value for Divisor
```

```
        btfsc      Dividend, 6       ;  Have we moved Divisor all the way
                                        up?
        goto       DivLoop

        bcf        STATUS, C         ;  Else, start shifting the values
        rlf        Count
        rlf        CountHI
        rlf        Divisor
        rlf        DivisorHI

        goto       StartLoop

DivLoop                              ;  Count and Divisor set up, start dividing
        movf       DivisorHI, w      ;  If Divisor < Dividend, sub Divisor
        subwf      DividendHI, w
        movwf      Temp
        movf       Divisor, w
        subwf      Dividend, w
        btfss      STATUS, C
        decf       Temp
        btfsc      Temp, 7           ;  Do we have a negative?
        goto       DivSkip           ;  No, skip over

        movwf      Divisor           ;  Else, it is, save new Dividend
        movf       Temp, w
        movwf      DivisorHI

        movf       CountHI, w        ;  Add the set bit to Quotient
        addwf      QuotientHI
        movf       Count, w
        addwf      Quotient

DivSkip                              ;  Finished, shift Divisor and Count
        bcf        STATUS, C
        rrf        DivisorHI
        rrf        Divisor

        rrf        CountHI           ;  Now, see if Count is finished
        rrf        Count

        btfsc      STATUS, C         ;  Is the Carry flag set?
        goto       DivLoop           ;  No, loop again

        movf       DividendHI, w     ;  Save the remainder
        movwf      RemainderHI
        movf       Dividend, w
        movwf      Remainder
```

C

Reuse, return, recycle

As I said in chapter 6, the best programmers get snippets of code that they like and continually use over and over again.

To try to make it easier for you, I've included a few routines and include files that I've developed that might make your life easier. I would be surprised if you can use everything in this appendix without change, but maybe there'll be things in here that you can use.

Useful routines

I have developed some routines that I use repeatedly to carry out standard functions. When using these routines, make sure you understand what temporary registers are used along with ports and bits within the ports. These routines can be put into subroutines or copied directly into the body of a program.

The first routine shifts out 8 bits to a clocked device:

```
ShiftOut                        ; Shift the contents of "w" out to
   movwf      TempByte          ;   the clocked register on PortB
   movlw      8                 ; 8 bits to shift out
   movwf      Counter
SO_Loop                         ; Come back here for each bit
   rrf        TempByte          ; Load "c" with least sig. bit
   btfsc      STATUS, C         ; Is the bit set?
   goto       SO_Set            ; Yes, set it when you shift out
   bcf        PORTB, Data       ; No, clear the shift out bit
   goto       SO_Toggle
SO_Set                          ; Set the bit to shift out
   bsf        PORTB, Data
SO_Toggle                       ; Toggle the clock line
   bcf        PORTB, Clock
   bsf        PORTB, Clock
   decfsz     Counter           ; Have we done all 8 bits?
   goto       SO_Loop           ; Nope, do the next One
   return                       ; ShiftOut end
```

Two of the more interesting things I have come up with are the "bcw" and "bsw" macros. These two macros mimic the "bcf" and "bsf" instructions by just resetting or setting a bit in the "w" register:

```
bcw MACRO bit
  addlw 0x0FF ^ ( 1 << bit )
 ENDM
bsw MACRO bit
  iorlw 1 << bit
 ENDM
```

"1" is shifted to the left "bit" times, making 1 << bit equivalent to 2 ** bit. Note that these macros will change the contents of the STATUS Zero flag (the "bcf"/"bsf" instructions don't affect any STATUS flags).

One routine I have come up with is the bubble sort for the PIC. The input is a single-dimensional array of values, and the output is a single dimensional array of the values sorted in ascending order. Table C-1 lists the variables used.

Table C-1. The variables used in the bubble sort

Variable	Description
reg1	Start of array of values to be sorted
rega	Array of sorted values
next	Location to put the next sorted value
llow	Value of the last lowest number
addr	Location of the last lowest number
lend	Location of the list end

The code is shown in Fig. C-1 and is designed for sorting four values. The size of the array to be sorted can be increased easily by changing the "lend" value at the start of the routine.

C-1 The code for the bubble sort.

```
Sort
;  Now, in the sorting routine
   movlw    rega              ;  Set up where you are storing the result
   movwf    next
   movlw    reg4              ;  For shrinking list, get the last addr
   movwf    lend              ;  Watch for the ending value

Loop                          ;  Loop around here until list is empty
   movlw    reg1              ;  Load FSR for searching for the lowest
   movwf    FSR
   movwf    addr              ;  At start, assume the first is lowest

   movf     INDF, w           ;  Get the current and use as the lowest
   movwf    llow              ;  Save it as the current lowest

Loop2                         ;  Loop here until fsr = lend
   movf     FSR, w            ;  Are we at the end?
   subwf    lend, w
```

```
     btfsc   STATUS, Z        ; If Zero flag is set, we're at the end
     goto    Save             ; Save the currently lowest value

     incf    FSR              ; Now, look at the next value

     movf    llow, w
     subwf   INDF, w          ; Do we have something that's lower?
     btfsc   STATUS, C        ; If Carry is set, then current is lowest
     goto    Loop2            ; If Carry set, then we still have lowest

     movf    INDF, w          ; Else, current is the lowest—save it
     movwf   llow
     movf    FSR, w           ; And, save the address it's at
     movwf   addr
     goto    Loop2            ; Loop around and look at the next

; The list has been checked and "low" and "addr" have lowest current
;    value and its address, respectively.
Save                          ; Now, save the currently lowest value
     movf    next, w          ; Store it in the FSR
     movwf   FSR

     movf    llow, w          ; Get the lowest
     movwf   INDF             ; Store it in the sorted List

     movf    next, w          ; Are we at the end of the list?
     sublw   regd
     btfsc   STATUS, Z        ; If NOT zero, then loop around
     goto    PEnd             ; Else, they match, end the program
     incf    next             ; Increment the pointer to the next
                              ; value

; The lowest current value has been put in the "sorted" list.
;    Now, shorten the list at the value we took out
     movf    addr, w          ; Get address the value was taken out
                              ; of
     movwf   FSR              ; Put in the FSR for later

Loop3                         ; Now, loop around storing the new list
     movf    FSR, w           ; Are we at the end of the list?
     subwf   lend, w
     btfsc   STATUS, Z        ; Is the Zero flag set?
     goto    Skip             ; Yes, list has been copied

     incf    FSR              ; Get the next value and store in current
     movf    INDF, w
     decf    FSR
     movwf   INDF
     incf    FSR              ; Increment the index and loop around
     goto    Loop3

Skip
     decf    lend             ; Decrement the ending spot
     goto    Loop

; Sort is all finished. Now, just loop around...
PEnd                          ; Program end
     return
```

There's a bit of a story to this routine. On the PICLIST, somebody asked for a routine that sorted four numbers in one list and put them in another. After working for about three hours, I came up with the solution in Fig. C-1.

It was quite an eye opener when I saw what other people came up with; whereas I improved the baseline code by about a three times (three times shorter execution time and about a third of the original code size), the best solutions were almost a hundred times better!

The solutions presented were designed to do exactly what was required: sort four numbers from one list and put them in another. Developing a macro for comparing two values and putting the lowest first, the solution was:

```
least          regc, regd        ;  Move the lowest value to front of
least          regb, regc        ;    beginning of list
least          regb, rega
least          regc, regd        ;  Move 2nd lowest value to 2nd from
least          regb, regc        ;    the front
least          regc, regd        ;  Put the two highest in order
```

The macro used to accomplish this was:

```
least          macro             reg1, reg2
   movf        reg2, w
   subwf       reg1, w
   btfsc       STATUS, C         ;  If no carry, swap the values
      goto     $+6               ;  Else, skip over the rest
   movf        reg1, w           ;  Now, swap the values
   xorwf       reg2, w
   xorwf       reg2
   xorwf       reg2, w
   movwf       reg1
   endm
```

The lesson I learned in all this was: Understand what the customer wants. While the routine I created was very clever and took a bit of work, it was not what the customer wanted.

Nothing pays off like reading the datasheets. Some time ago, the question was asked on the PICLIST regarding what was the easiest way of determining what type of PIC certain code was running on. The question was asked in relation to a piece of code written for the two devices—the 16C61 and the 16C84—and how the software could determine which device it is actually running on.

After a cursory reading of the datasheets, you will see that the only real difference between the 16C61 and 16C84 is the data memory. The 'C61 doesn't have any, and the 'C84 has nonvolatile (i.e., the contents don't change when power is removed) EEPROM data locations. The initial response to the question was to write a value into EEPROM. If the value programmed into the EEPROM could be read back, then the code was running on a 16C84.

The much more elegant solution relied on more than a cursory inspection of the code along with an eye of what the end objective was. The end objective was trying to come up with a method of determining whether or not the device the code was running on had any EEPROM.

To do this, the register set of the two devices was compared and the major difference was found in the EEPROM area of the 'C84 (the 'C61 doesn't have any comparable registers). Looking at the registers and the various bits, it was discovered that the "WREN" bit in the EECON1 register is reset upon power-up and can be written to and read from. In the 'C61, the same register (address 0x088) doesn't exist and is always read as a zero.

This lead to the simple code that, when finished, returns 61 or 84 in "w," depending on the PIC it is running on:

```
movlw      61                 ;  Assume that we are running with a 'C61
bsf        STATUS, RP0        ;  Jump to bank 1
bsf        EECON1, WREN       ;  Set the "WREN" bit
btfsc      EECON1, WREN       ;  If "WREN" bit is reset, jump past next
 movlw     84                 ;  We actually have a 'C84
bcf        STATUS, RP0        ;  Return to bank 0
```

This code only requires six addresses and six cycles, which is considerably less than actually writing to an EEPROM location.

Another useful routine in the PIC is being able to have a (reasonably) accurate delay that does not require a significant amount of PIC resources.

Delays in the PIC can be done in two ways. The first (and something that's discussed throughout the text) is using a timer and an interrupt for timing (waiting for) events. The second is to use a polling loop for waiting for an event.

The simplest loop I have found is:

```
movlw      Value              ;  Enter the count value
movwf      Count
Dlay                          ;  Loop until Count + (Counthi * 256) = = 0
 decfsz    Count              ;  Decrement the low value of the number
 goto      Dlay
```

To calculate the time required for the loop, we know that the "decfsz" instruction takes one cycle and the "goto" instruction takes two. This means that, for most loops, the number of cycles required is 3 (for the loop over, we add 3 more cycles). Each instruction cycle requires 4 clock cycles in which to execute, so every iteration of the "Dlay/decfsz Count/goto Dlay" loop actual takes 12 clock cycles.

Therefore, the time delay for this loop is:

$$time\_delay = \frac{\left(\left(\dfrac{4 \text{ cycles} \times 4}{\text{cycles}}\right) + \left(\left(\dfrac{3 \text{ cycles} \times 4}{\text{cycles}}\right) \times (Count - 1)\right)\right)}{freq}$$

The four cycles at the start of the equation is the "movlw Value/movwf Count/decfsz Count" (for Count equals 1) and the three cycles is for the "decfsz Count/goto Dlay" combination.

This formula can be scaled up as shown here:

```
movlw      Low_Value          ;  Enter the count value
movwf      Count
movlw      High_Value         ;  High value
movwf      Counthi
Dlay                          ;  Loop until Count + (Counthi * 256) == 0
 decfsz    Count              ;  Decrement the low value of the number
 goto      Dlay
 decfsz    Counthi            ;  Decrement high value of the number
 goto      Dlay
```

The delay is specified by the formula:

$$time\_delay = \frac{\left(\left(\dfrac{6 \text{ cycles} \times 4}{\text{cycles}}\right) + \left(\left(\dfrac{3 \text{ cycles} \times 4}{\text{cycles}}\right) \times (Count - 1)\right)\right)}{freq}$$

$$+ \frac{\left(\dfrac{2 \text{ cycles} \times 4}{\text{cycles}}\right)}{freq}$$

$$+ \frac{((2 \text{ cycles} + (3 \text{ cycles} \times 255) + 3 \text{ cycles}) \times \dfrac{4}{\text{cycles}} \times (Counthi \times 1))}{freq}$$

This is a very complex equation, that I usually simplify it to:

$$time\_delay = \frac{(3 \text{ cycles} \times 256) \times \dfrac{4}{\text{cycles}} \times (Counthi \times 2)}{freq}$$

By incrementing "Counthi" (which allows me to ignore "Count"), over the general case, I make sure I get at least the target delay. Generally, when delays require a 16-bit number, the loss of accuracy by the simplification isn't a major problem.

For example, if I wanted a 5-msec delay in a 4-MHz PIC, The simplified formula would yield:

$$Counthi = \left(\frac{freq \times time\_delay}{(3 \times 256 \times 4)}\right) + 2$$

$$= \left(\frac{4 \text{ MHz} \times 5 \text{ msec}}{3072}\right) + 2$$

$$= 8$$

Put through the previous formula (assuming "Count" was equal to one), this works out to:

$$time\_delay = \frac{(8 + (2 + 765 + 3) \times 4 \times 7)}{4 \text{ MHz}}$$

$$= 5.39 \text{ msec}$$

which is approximately 8% above the 5 msec target.

Timing an event can be done using the "incfsz" instruction:

```
      clrf        Count              ;  Clear the counter values
      clrf        Counthi
.
.
                                     ;  Wait for setup of an event
.
.
Loop                                 ;  Return here for each iteration
      incfsz      Count              ;  Increment the 16-bit counter values
      incf        Counthi
      btfsc       REGn, Bitx         ;  Wait for the event to complete
      goto        Loop
      movf        Counthi, w         ;  Get actual high value byte of Count
```

```
subwf          Count, w
movwf          Counthi
```

This method will provide a 16-bit count with granularity down to 5 instruction cycles (20 clock cycles). Greater timing accuracy (smaller granularity) can be accomplished by running TMR0 with no prescaler input and a simple test loop:

```
btfsc REGn, Bitx              ; Wait for the event to complete
  goto $ - 1                  ; Loop back until it does
```

There are times when you might want to measure the frequency of a signal (e.g., create a tachometer for a motor). The frequency is simply the number of times an event happens in a given period of time.

This can be done very easily with the PIC using a foreground loop of known duration and the TMR0 running in the background, counting the events (or pulses).

For example, let's say you wanted to measure the number of pulses that happen in 1 msec. This can be done easily using a 16C84 running at 4 MHz (to get a 1 μsec instruction cycle rate). The pulses can be input directly into TMR0 through the RA4/TOCK1 pin. TMR0 will be incremented every two pulses from RA4/TOCK1, if the prescaler is not selected for TMR0.

To count the number of pulses in 1 msec, the following code would be used:

```
bsf            STATUS, RP0
movlw          0x0F8                 ; Set up OPTION register so TMR0
movwf          OPTION_REG & 0x07F    ;  will use RA4 as a source WITHOUT
bcf            STATUS, RP0           ;  the prescaler
movlw          73                    ; Set up forground counter for 1000
movwf          Count                 ;  instruction cycles
movlw          2
movlw          Counthi
bcf            INTCON, T0IF          ; Clear the TMR0 overflow flag
clrf           TMR0                  ; Start TMR0 from 0
decfsz         Count                 ; Wait 1 msec with TMR0 running
  goto         $ - 1
decfsz         Counthi
  goto         $ - 3
movf           TMR0, w               ; Get the 1/2 the number of pulses
                                     ;  in 1 msec
```

This code only requires two file registers ("Count" and "Counthi") for the 16-bit foreground delay.

The basic operation of this code is to use TMR0 to count the number of incoming pulses while the foreground delay operation waits 1000 cycles (at 4 MHz, 1 cycle is 1 μsec long) to get a 1-msec wait.

TMR0 goes through a synchronization step, which means that only half of the cycles are counted. Therefore, this code will count up to 510 events in 1 msec, rather than the 255 expected.

Why is 255 expected? If the TMR0 rolls over during the loop, there's a chance that some data will be lost. Therefore, I recommend clearing the Timer Overflow bit (T0IF) and checking it once the TMR0 contents have been stored in "w." If the Overflow bit is set, then you can't be sure of the contents of TMR0 (i.e., did it overflow once, twice, or...). When this happens, you should reset your delay loop to half the value and repeat the sampling or using the prescaler. This would be continued until you have a value where the Overflow isn't set.

While this block of code looks pretty good for finding out how many samples take place in a given period of time, it should be noted that, at best, you can only expect an accuracy of the most significant five or six bits of TMR0 (i.e., the hundreds and the tens). I wouldn't count on this code to be any more accurate than that. The accuracy is at its best by planning the mainline delay loop such that you get a TMR0 value between 100 and 200 (decimal).

The advantages of this code is the speed in which it is able to sample with a fair degree of accuracy. If extreme accuracy is required, other methods and hardware should be used.

Include files

The three files provide some useful functions that I felt were not available from the currently available Microchip Application Notes and code available on the various other PIC resources. While these include files have been tested and do work, you might encounter problems because of timing or hardware variances. This means that some debugging might be required to get them to run properly in a specific application.

After creating the "Frosty the Snowman" Christmas decoration, I felt that the code used to create the music would be useful in other situations. I stripped out the LFSR code and just created the "TuneMkr.zip" package, which includes the necessary source code to play a tune as well as some PC utilities for converting the music into a format the source code can use. "TuneMkr.Zip" can be unloaded using PKUNZIP 2.04G.

To simplify the creation of the program, the two programs "Notes" and "MusicMkr" are used (this is explained in the "Readme.101" and "TuneMkr.bat" files) to create "Sound.Fyl" and "Notes.Fyl," which are included in "Sound.Asm" at MPASM assembly time.

Looking back over the code, I see that this application was created before I really understood and thought about creating compact, understandable source code. If I were to do it again, I would make the following changes:

- Develop the Assembler code independently and not use my compiler to do the job. The code in Sound.Asm as it stands is pretty hard to understand.
- I would use the "CBLOCK" and "DT" Assembler commands to make the created registers more easily reallocated and the source code more compact.
- I would design it for different frequency PIC clocks. While it does work well with the default 1-MHz clock, a higher frequency clock would give a greater dynamic range.
- Rather than limiting the song size to 81 notes by using a simple table organization, I would probably change the table code to allow a table that goes over a 256-address boundary.
- The music, as provided, doesn't allow any way of inserting pauses.

If, after reading this list, you decide to modify the package to fix the problems I just listed, let me know and I'll make sure it's available to other people.

An LCD can be an invaluable debugging tool. I didn't realize how good it can be until I began to write the serial-LCD interface code. The only problem with using an LCD is the amount of pins required (at least six). This was a concern with the servo controller, so I came up with the code in Fig. C-2, which, by using a serial in-parallel out shift register will give you a 3-wire LCD interface.

C-2 Code for a 3-wire LCD interface.

```
  subtitle "LCD3LINE - Control an LCD using 3 lines."
;
;   These routines will initialize, send instructions, and send
;     characters to a standard Hitachi 44780 controlled LCD. A 3-line
;     interface is used to reduce the number of PIC pins used on the LCD.
;
;   NOTE:  Before the initialization is to be executed by the PIC, 15
;     msecs must elapse before the "Init" subroutine is called.
;
;   The hardware is:
;
;                                          Gnd ----| 1    44780
;                                          Vcc ----| 2    Controlled
;                                  Contrast V ----| 3    LCD
;                                      +-----------| 4
;                                      |  Gnd ----| 5
;                  +------------------ | ----------| 6
;                  | |                 |           | 7
;                  | |     -------      |          | 8
;        RB7 |-----+   | Shift |-----+  |          | 9
;                  |   | Reg   |--------------+ | 10
;                  |   |       |-----------+  +--| 11
;        RB6 |-----------| Clk   |-------+  +------| 12
;        RB5 |-----------|_D____|---+  +---------| 13
;                  |                   +-------------| 14
;
;   Hardware notes:
;     The PIC speed is immaterial
;     RB5 is connected to Data Out to the LCD Control
;     RB6 is connected to the LCD shift register clock
;     RB7 is connected to 'E' bit of the LCD
;     The shift register can be any 5-bit (or more) serial
;     in/parallel out
;       shift register
;     LCDTemp, LCDNTemp, LCDCount  - file registers used for data transfer
;
;   External routines required:
;     Dlay5ms - Wait 5 ms
;     Dlay160 - Wait 160 usec for the instruction to complete
;
;   Myke Predko
;   97.01.23
;

LCDInit                        ;  Initialize the LCD

    movlw  3                   ;  Start with the initialization
    bcf    PORTB, 5            ;  Make sure the RS flag = 0
    call   LCDNybble           ;  Output the nybble
```

C-2 Code for a 3-wire LCD interface. *Continued*

```
        call    Dlay5ms             ; Wait for the LCD to power up

        movlw   3                   ; Continue with the reset
        bcf     PORTB, 5
        call    LCDNybble
        call    Dlay160             ; Wait 160 usec

        movlw   3                   ; Reset again
        bcf     PORTB, 5
        call    LCDNybble
        call    Dlay160

        movlw   2                   ; Now, set interface length to 4 bits
        bcf     PORTB, 5
        call    LCDNybble
        call    Dlay160

        movlw   0x028
        call    SendINS             ; Now, can just send instructions—set
                                    ;   4 bits, 2 lines
        movlw   0x00C               ; Turn display on
        call    SendINS
        movlw   0x001               ; Clear the display, set up the cursor
        call    SendINS
        movlw   0x006               ; Set the entry mode

        return


LCDINS                              ; Send the byte as an instruction

        movwf   LCDTemp

        bcf     INTCON, GIE

        bcf     PORTB, 5            ; Start with the R/S line low

        swapf   LCDTemp, w         ; Get the high nybble to send
        call    LCDNybble

        bcf     PORTB, 5           ; Now, send the low nybble

        movf    LCDTemp, w
        call    LCDNybble

        bsf     INTCON, GIE

        call    Dlay160            ; Wait for the LCD to process instruct

        movlw   0x0FC              ; "Clear Display" and "Cursor At Home"
        andwf   LCDTemp, w         ;   require 5 msec delay to complete
        btfsc   STATUS, Z
        call    Dlay5ms
```

```
        return

    LCDCHAR                              ;  Send the byte as character data

        movwf   LCDTemp                  ;  Save the value for the second nybble

        bcf     INTCON, GIE              ;  Turn OFF interrupts during write

        bsf     PORTB, 5                 ;  Start with the R/S line low

        swapf   LCDTemp, w               ;  Get the high nybble to send
        call    LCDNybble

        bsf     PORTB, 5                 ;  Now, send the low nybble

        movf    LCDTemp, w
        call    LCDNybble

        bsf     INTCON, GIE

        call    Dlay160                  ;  Wait for LCD to display the character

        return

    LCDNybble                            ;  Send the nybble (with what's set in
                                         ;    PORTB.5 to begin)

        movwf   LCDNTemp                 ;  Save the 4 bits to display

        bcf     PORTB, 6                 ;  Clock out the contents of PORTB.5
        bsf     PORTB, 6

        movlw   4                        ;  Now, shift out 4 more bits
        movwf   LCDCount

    LNLoop                               ;  Loop around here

        bcf     PORTB, 5
        rrf     LCDNTemp                 ;  Load carry with the LSB
        btfsc   STATUS, C
        bsf     PORTB, 5                 ;  Set the bit (if appropriate)

        bcf     PORTB, 6                 ;  Clock out the bit
        bsf     PORTB, 6

        decfsz  LCDCount
        goto    LNLoop

        bsf     PORTB, 7                 ;  Toggle the 'E' clock
        bcf     PORTB, 7

        return
```

Along with the LCD Interface, I²C serial devices can be very useful in storing and retrieving collected data (or programs, as is done in the emulator or servo controller). The routines in Fig. C-3 use the open-collector pin (RA4) on mid-range PICs to interface with an I²C device as a single bus master.

C-3 Routines to interface with an I²C device as a single bus master.

```
  subtitle  "I2C - I2C Master routines using RA3/RA4."
;
;   The routines contained in this .inc file allow communication between
;    a PIC equipped with RA4 as an "open collector" output. This code is
;    designed for the PIC to be the only master on the I2C bus.
;
;   This code is designed to work for 7 bit addressed I2C devices. The
;    PIC speed is nominally 4 MHz. For faster PIC clock speeds, a lower
;    pull-up value should be used.
;
;   At the start of the host software, RA4 and RA3 must be set to
;    "Output" (their respective TRISB bits set to "0"). The normal
;    inactive state consists of the SDA and SCL lines low.
;
;   During transmit, the acknowledge lines are NOT checked to see if the
;    receiver is active and accepting the incoming data.
;
;   A typical I2C EEPROM Read or two bytes consists of:
;
;    call  I2CStart           ;  Send the start bit
;
;    movlw 0x005              ;  Send the EEPROM device enable
;    call  I2CSend
;
;    movf  Address, w         ;  Send the address within the EEPROM
;    call  I2CSend
;
;    call  I2CStart           ;  Now, going to start reading
;
;    movlw 0x085              ;  Enable the EEPROM read
;    call  I2CSend
;
;    clrw                     ;  Receive the 1st byte (acknowledge)
;    call  I2CReceive
;    movwf Data1st
;
;    movlw 1                  ;  Receive the 2nd byte - NO ack
;    call  I2CReceive
;    movwf Data2nd
;
;    call  I2CStop            ;  Send stop bit to stop the transfer
;
;
;  Hardware notes:
;   The PIC is the only master on the I2C bus
;   RA3 is connected to "SCL" on the I2C bus
```

```
;    RA4 is connected to "SDA" on the I2C bus (which has 1K to 10K pull-up)
;    I2CTemp is a file register
;
;    Myke Predko
;    97.02.18
;

I2CStart                              ;  Send the start bit on the I2C bus

    bsf     PORTA, 4                  ;  Make the data line high
    goto    $ + 1
    bcf     PORTA, 3                  ;  Make the clock line low
    nop
    bsf     PORTA, 3                  ;  Set the clock high
    goto    $ + 1                     ;    for 5 clock cycles
    goto    $ + 1
    nop
    bcf     PORTA, 4                  ;  Drop clock line to go into start mode
    goto    $ + 1                     ;    for 5 clock cycles
    goto    $ + 1
    nop
    bcf     PORTA, 3                  ;  Start clock train
    goto    $ + 1

    return


I2CStop                               ;  Send the stop bit to the I2C bus

    bcf     PORTA, 4                  ;  Data line low
    goto    $ + 1
    goto    $ + 1
    bsf     PORTA, 3                  ;  Set the clock line high
    goto    $ + 1
    goto    $ + 1
    bsf     PORTA, 4                  ;  Data goes high while
    goto    $ + 1                     ;    clock is high for stop bit
    bcf     PORTA, 3                  ;  Set clock low again
    goto    $ + 1
    nop

    return


I2CSend                               ;  Send the data out

    movwf   I2CTemp                   ;  Store the value to send

    movlw   8                         ;  Sending 8 bits

I2CWLoop                              ;  Loop around here until each bit sent

    rrf     I2CTemp                   ;  Figure out what we have
    btfsc   STATUS, C                 ;  Is the LSB set?
     goto   I2CWHIGH                  ;  Yes...
```

C-3 Routines to interface with an I²C device as a single bus master. *Continued*

```
        nop
        bcf     PORTA, 4            ;  Drop the data line

        goto    I2CWStrobe         ;  Now, strobe the line

I2CWHIGH                           ;  Make the line high

        bsf     PORTA, 4

        goto    I2CWStrobe

I2CWStrobe                         ;  Strobe the clock line

        bsf     PORTA, 3           ;  Make the line high for 4 usec
        goto    $ + 1
        goto    $ + 1
        bcf     PORTA, 3

        addlw   0 - 1              ;  Take the number down by one
        btfss   STATUS, Z          ;  Stop at zero
         goto   I2CWLoop

        goto    $ + 1              ;  Now, receive the ack

        bsf     PORTA, 4           ;  Make sure data bus is high for ack

        goto    $ + 1
        movf    PORTA, w           ;  Strobe clock (nothing else changed)
        iorlw   0x008
        movwf   PORTA
        goto    $ + 1              ;  NOTE: Ack line is not checked
        goto    $ + 1
        andlw   0x0F7
        movwf   PORTA              ;  And, bring the line low

        return                     ;  Return to the caller


I2CReceive                         ;  Receive a data byte

        movwf   I2CTemp            ;  Save the ack bit in "w"

        movlw   8                  ;  Sending 8 bits

        rrf     I2CTemp            ;  Load the Carry flag with the ack bit

I2CRLoop                           ;  Loop around here until each bit read

        rrf     I2CTemp            ;  Shift the data down

        iorlw   0                  ;  Have we done all 8 bits?
        btfsc   STATUS, Z
         goto   I2CR_Ack           ;  Yes, Carry has the ack bit
```

```
        addlw   -1                      ; Decrement the count

        bsf     PORTA, 4                ; RA4 is controlled by I2C device
        bsf     PORTA, 3                ;   Strobe the clock

        nop                             ; Keep constant timing

        bcf     STATUS, C               ; Carry to be used for the data
        btfsc   PORTA, 4                ; Skip the Carry set if the line is low
         bsf    STATUS, C

        bcf     PORTA, 3                ; Drop the clock line

        goto    I2CRLoop                ; Shift carry in and check next value

I2CR_Ack                                ; Now, send specified acknowledge bit

        bsf     PORTA, 4                ; Assume there's no ack
        btfss   STATUS, C
         bcf    PORTA, 4                ;   Or, there is

        bsf     PORTA, 3                ; Strobe out the acknowledge

        goto    $ + 1
        goto    $ + 1

        bcf     PORTA, 3

        return                          ; Return to the caller
```

The last include file (Fig. C-4) is an interrupt-driven serial receiver and matching transmitter. The receiver waits for a transition on the RB0 line, checks for a valid start bit, then reads in the data and saves it if there is a valid stop bit. Along with the start/stop bit checking, this routine also puts the received characters into a circular buffer, which allows the mainline routine to process for a while without having to access the received serial buffer.

C-4 An interrupt-driven serial receiver and matching transmitter.

```
 subtitle   "Serial serial 9600 bps interrupt Rx/Tx."
;
; These routines are used to provide a serial (digital logic)
;   from a PIC. These routines do not use the serial communications
;   hardware in some PICs and do require interrupts. Therefore,
;   The 16C61, 16C71, and 16C84 are the recommended target devices.
;
; Note that the input is DIGITAL LOGIC. For RS-232 "cheats", the
;   values read/output will have to be inverted.
;
; Currently they are set up for a PIC running at
;   3.579545 MHz (color burst frequency) and 9600 bps. This
```

C-4 An interrupt-driven serial receiver and matching transmitter. *Continued*

```
;     can be changed by changing the count value.
;
;     Note that the received data is put into the 5-byte circular
;       "buffer".
;
;     It is assumed that the host the PIC running this code can use
;       a simple 3-wire interface.
;
;     Hardware notes:
;       The PIC is running with a 3.579545 MHz (color burst) crystal for
;       16C84
;       RA1 is connected to RS-232 "TX" line
;       RB0 is connected to RS-232 "RX" line
;       NOTE:  RA1 is used to transmit data rather than RA0 because
;         RA0 is usually used to look like the LSB of PORTB.
;
;     Myke Predko
;     96.09.22


;   Declarations
  cblock _VarEnd
BitCount                          ;   Count value for the bits
Char                              ;   Temporary storage of the
character read in
Temp                              ;   Temporary value for Program
RRCount                           ;   Loop count for reading the
characters
SCCount
Next                              ;   Pointer to the next value in the
chain
ShadowFSR                         ;   Saved the FSR buffer value
_w                                ;   Context save registers
_status
BufferStart                       ;   Character buffer
Buffer1
Buffer2
Buffer3
BufferEnd
  endc


;   Constants
Bit      EQU     27               ;   Delay between the bits, using the
BitStart EQU     Bit / 3          ;   delay to check in character middle

;   The bit count is calculated by the formula:
;
;     Period = 1 / bps             ;   Get the period required.
;     Total Cycles = ( Period * ( PIC_Frequency / 4 )) - 15
;                                  ;   Number of cycles to execute
;     Bit = ( Total_Cycles / 3 ) + 1
;                                  ;   The actual number of Dlay loops
```

```
;
;   For example: Getting 2400 bps at 4 MHz:
;     Period = 1 / 2400 = 0.0004167
;     Total Cycles = ( 4.167(10^-4) * 4(10^6) / 4 ) - 15 = 402
;     Bit = ( 402 / 3 ) + 1 = 135


;   RX interrupt handler
;   Read serial routine. When character read and put in buffer, return
RXInt

    movwf  _w                 ;  Save the context registers
    movf   STATUS, w
    movwf  _status
    bcf    STATUS, RP0        ;  Make sure we're in bank 1

ReadRS232                     ;  Start of original serial read routine

    decfsz BitCount           ;  Wait for halfway through the char
     goto  $ - 1

    btfss  PORTB, 0           ;  Do we *still* have the start bit?
     goto  ShortRRLoop        ;  Point of inversion depending on logic

    movlw  0x010              ;  Reset interrupts before returning
    movwf  INTCON
    goto   RREnd              ;  Wait for the next transition

ShortRRLoop                   ;  Now, loop around here for each char

    movlw  Bit + 2            ;  Load and wait for the bit count
    movwf  BitCount           ;     NOTE: 3 goto $ + 1 removed
    decfsz BitCount           ;  Now Dlay
     goto  $ - 1

    rrf    PORTB, w           ;  Get the bit value coming in
    rrf    Char               ;  Update the character

    decfsz RRCount            ;  Do 8x
     goto  ShortRRLoop

    movlw  Bit                ;  Now, look at the stop bit
    movwf  BitCount
    decfsz BitCount
     goto  $ - 1

    movlw  0x010              ;  Reset the interrupt enable (in case
    movwf  INTCON             ;     one comes in during code below)

    btfss  PORTB, 0           ;  Check to see if stop bit is there
     goto  RREnd              ;  It's not, ignore the character

HaveRS232                     ;  Have correct RS232 value - end/return

    incf   Next               ;  Now, store the character in buffer
```

C-4 An interrupt-driven serial receiver and matching transmitter. *Continued*

```
        movlw   BufferEnd + 1           ;  Are we at the end of the buffer?
        subwf   Next, w
        btfss   STATUS, C               ;  If Carry set, no
         movlw  BufferStart - ( BufferEnd + 1 )  ;  Else, have to roll to start
        addlw   BufferEnd + 1           ;  Reset the start of next

        xorwf   FSR, w                  ;  Swap next with FSR
        xorwf   FSR                     ;    (Shadow contains current FSR)
        xorwf   FSR, w
        movwf   Next                    ;  Save the old FSR value

        movf    Char, w                 ;  Store the character in the buffer
        movwf   INDF

        movf    Next, w                 ;  Restore the FSR
        xorwf   FSR, w
        xorwf   FSR
        xorwf   FSR, w
        movwf   Next                    ;  Store the current next value

RREnd                                   ;  Return to where routine was called

        movlw   8                       ;  Reload the bit counter for the loop
        movwf   RRCount

        movlw   BitStart                ;  Reset counters for next serial char
        movwf   BitCount

IntEnd                                  ;  Finished with the interrupt

        movf    _status, w              ;  Restore context registers and return
        movwf   STATUS
        swapf   _w
        swapf   _w, w

        retfie


;   Interrupt setup routine - Init the variables for the RXInt handler
;   NOTE:  It is assumed that PORTA.1 is set for output
RXSetup

        movlw   BitStart                ;  Set up the values for the interrupt
        movwf   BitCount
        movlw   8                       ;  Want to read 8 characters
        movwf   RRCount

        movlw   BufferStart             ;  Initialize the indexes to the buffer
        movwf   ShadowFSR
        movwf   Next

        movlw   0x090                   ;  Turn on the receive interrupts
        movwf   INTCON
```

```
        return


;  Character read routine. Wait for something to be available
GetCHAR

        movf    ShadowFSR, w        ;  Get the last character read
        subwf   Next, w             ;   in the buffer and, if it
        btfss   STATUS, Z           ;   hasn't been read, return it
         goto   GetCHAR             ;  Else, wait for the next

        incf    ShadowFSR, w        ;  Point to the next value in
        addlw   0 - ( BufferEnd + 1 );   the table
        btfsc   STATUS, C
         addlw  BufferStart - ( BufferEnd + 1 )
        addlw   BufferEnd + 1
        movwf   ShadowFSR           ;  Save it for the next read

        movwf   FSR                 ;  Get the newly entered character
        movf    INDF, w

        return


;  Character send routine - Send the character in "w"
SendCHAR                            ;  Serial send the character to the host

        movwf   Temp                ;  Save the character to send

        movlw   8                   ;  Set up the 8 bits to send
        movwf   SCCount

        bcf     PORTA, 1            ;  Output the start bit

SCLoop                              ;  Loop here for each character

        movlw   Bit                 ;  Delay the length of the bit
        movwf   Count
        decfsz  Count
         goto   $ - 1

        rrf     Temp                ;  Rotate the bit over
        btfsc   STATUS, C           ;  Do we have to Send a '1'?
         goto   SCOne

        nop                         ;  Send a zero
        bcf     PORTA, 1            ;  NOTE: It is inverted!

        goto    SCLoopEnd

SCOne                               ;  Send a one

        bsf     PORTA, 1            ;  NOTE: It is inverted!

        goto    $ + 1               ;  Delay 2 cycles to even up
```

C-4 An interrupt-driven serial receiver and matching transmitter. *Continued*

```
SCLoopEnd                       ;  Do this for 8 bits

  nop                           ;  To time everything out

  decfsz SCCount
   goto  SCLoop

  nop                           ;  Keep everything on track

  movlw  Bit                    ;  Delay for the bit
  movwf  Count
  decfsz Count
   goto  $ - 1

  goto   $ + 1                  ;  Delay the correct number of cycles
  goto   $ + 1

  bsf    PORTA, 1               ;  Send a zero stop bit

  nop
  movlw  Bit                    ;  Wait number of cycles before return
  movwf  Count
  decfsz Count
   goto  $ - 1

  return
```

The integral transmit routine is admittedly nothing special, but it does use the same count/delay values as the receiver, which makes it natural to work with the receive routine.

As noted in the comments, this code must be modified to run with "positive" (TTL) logic (i.e., a "1" is at Vdd).

D

PIC resources

As the last part of the book, I wanted to leave you with a list of PIC resources that will help you as you develop your own applications for the PIC.

You'll notice that I have included a lot of Internet resources. The reason for this is because I believe the 'Net is the most efficient method of requesting and receiving technical information. This feeling is more than confirmed by the PIC and all the information that is available on the 'Net—both from various Web sites to the PICLIST.

Please note that these addresses (mail, e-mail, and Web) may change without notice.

Microchip

Microchip Technology Inc.
2355 West Chandler Blvd.
Chandler, AZ 85224-6199
Tel: (602)782-7200
Fax: (602)899-9210
Web: http://www.microchip.com

Contacting me

I can be contacted by sending a note to "emailme@myke.com" or visit my web site at:
http://www.myke.com

However, as I've said elsewhere, for technical questions or suggestions, contact me through the PICLIST so that the information can be made available to a much larger audience.

The PICLIST

The PICLIST is an Internet listserver that distributes PIC related e-mail to a sub-scriber list. To join the list, send an e-mail note to "listserv@mitvma.mit.edu" with the words "subscribe piclist" in the first line of the body (not the subject line).

To get off the list, send an e-mail note to "listserv@mitvma.mit.edu" with the words "unsubscribe piclist" in the first line of the body.

One note about the PICLIST: While it is a terrific technical resource, it can be overwhelming with the volume of mail it generates (often 100+ e-mail messages per day). Initially, you might want to monitor the PICLIST through list archives (given later in this appendix) rather than subscribe.

My favorite PIC Web sites

As I said earlier, a lot of the best resources for the PIC can be found on the Web. Here are my favorite eight Web sites with some notes on what you can find. You might find other sites that you like better. If you do, I would appreciate it if you would let me know.

Microchip's Web site

http://www.microchip.com
- PIC datasheets and application notes
- PIC FAQs
- Microchip serial EEPROM datasheets and FAQs
- PIC development tools (MPASM, MPSIM, MPLAB, etc.)
- Web resources
- Microchip distributors and manufacturer reps

Eric Smith's PIC Project page

http://www.brouhaha.com/~eric/pic
- Eric's projects
- "Tips and Tricks"
- PIC Web resources

Ormix's English home page

http://www.ormix.riga.lv/eng/index.htm
- PIC resource information
- Development (programmer) software
- Programmer schematics
- List of in-circuit emulators
- Kits and development boards
- Applications

Fast Forward's home page (administered by Andy Warren)

http://www.geocities.com/SiliconValley/2499
- The "Embedded Systems Answer Line," containing a list of Q&As about the PICs (and other microcontrollers)

David Tait's PIC resource page

http://www.man.ac.uk/~mbhstdj/piclinks.html
- David Tait's programmer and software information
- List of PIC resources and products

DonTronics home page (administered by Don McKenzie)

http://www.labyrinth.net.au/~donmck
- Don's meeting place for designers and programmers
- PIC information and product references
- SimmStick information

Wirz Electronics home page

http://www.wirz.com
- PICs and other electronics at bulk prices
- Source for PIC applications and projects (including the emulator and enhanced serial-LCD interface)

Parallax Inc. home page

http://www.parallaxinc.com
- PIC products and application notes
- Basic Stamp information and application notes
- Information on connecting to Parallax-run PIC and Stamp lists

Periodicals

Here are a number of magazines that give a lot of information and projects on PICs. Every month, each magazine has better than a 50% chance of presenting a PIC application.

Circuit Cellar Ink

P.O. Box 698
Holmes, PA 19043-9613
Tel: 1(800)269-6301
BBS: (860)871-1988
Web: http://www.circellar.com/

Gernsback Publications

Electronics Now
Popular Electronics

Subscription Department
P.O. Box 55115
Boulder, CO
Tel: 1(800)999-7139
Web: http://www.gernsback.com

Microcontroller Journal

Web: http://www.mcjournal.com/
This is published on the Web.

Nuts & Volts

430 Princeland Court
Corona, CA 91719
Tel: 1(800)-783-4624
Web: http://www.nutsvolts.com

Everyday Practical Electronics

EPE Subscriptions Dept.
Allen House, East Borough
Wimborne, Dorset
BH21 1PF
United Kingdom
Tel: +44 (0)1202 881749
Web: http://www.epemag.wimborne.co.uk

Web sites of interest

While none of these are PIC specific, they are a good source of ideas, information, and products which will make working with the PICs a bit easier and interesting.

Howard H. Sams and Company Internet Guide to the Electronics Industry

Web: http://pobox.com/~electronics/
Basic primer on the Internet as it relates to electronics as well as a directory to electronics related URLs (PICs included).

Seattle Robotics Society

Web: http://www.hhhh.org/srs/
The Seattle Robotics Society has lots of information on interfacing digital devices to such "real world" devices as motors, sensors, and servos. They also do a lot of neat things.

List Of Stamp Applications (L.O.S.A)

Web: http://www.hth.com/losa.htm
The List Of Stamp Applications will give you an idea of what can be done with the Basic Stamp (along with the PIC), especially if you have a cat that needs medication.

Adobe PDF Viewers

Web: http://www.adobe.com
Adobe .pdf file format is used for Microchip, Parallax, and most other datasheets and application notes.

Hardware FAQs

Web: http:paranoia.com/~filipg/HTML/LINK/LINK_IN.html
A set of FAQs (Frequently Asked Questions) about the PC and other hardware platforms that will come in useful when interfacing the PIC to a Host.

Suppliers

The following companies supplied components that are used in this book:

Digi-Key

Excellent Source for PICs and other electronic parts. Most orders received the next day.

Digi-Key Corporation
701 Brooks Avenue South
P.O. Box 677
Thief River Falls, MN 56701-0677
Tel: 1(800)DIGI-KEY
Fax: (218)681-3380
Web: http://www.digi-key.com/

AP Circuits

Prototype raw cards. Three-day service.

Alberta Printed Circuits Ltd.
#3, 1112-40th Avenue N.E.
Calgary, Alberta T2E 5T8
Tel: (403)250-3406
BBS: (403)291-9342
Web: http://www.apcircuits.com/
Email: staff@apcircuits.com

Wirz Electronics

Wirz Electronics is a full-service Microchip PIC component and development system supplier. Wirz Electronics was the main distributor for projects contained in this book (the Emu 18-pin PIC emulator and serial-LCD interface). Wirz Electronics also carries the SimmStick prototyping system, which was featured in the servo-controller project.

Wirz Electronics
P.O. Box 457
Littleton, MA 01460-0457
Tel: 1(888)BUY-WIRZ
Web: http://www.wirz.com/
Email: sales@wirz.com

Tower Hobbies

Excellent source for servos and R/C parts useful in home-built robots.

Tower Hobbies
P.O. Box 9078
Champaign, IL 61826-9078
Tel: (217)398-3636
Fax: (217)356-6608   1(800)637-7303
Ordering: 1(800)637-4989
Support: 1(800)637-6050
Web: http://www.towerhobbies.com/
Email: orders@towerhobbies.com

JDR

Components, PC parts/accessories, and hard-to-find connectors (like the 18-pin 0.3" IDC connector used on the Emu).

JDR Microdevices
1850 South 10th St.
San Jose, CA 95112-4108
Tel: (408)494-1400   1(800)538-500
Fax: 1(800)538-5005
BBS: (408)494-1430
Web: http://www.jdr.com/JDR
Email: techsupport@jdr.com
CompuServe: 70007,1561

Newark

Components, including the Dallas line of semiconductors (the DS275 is used for RS-232 level conversion in this book).

Tel: 1(800)4-NEWARK
Web: http://www.newark.com/

Mondo-Tronics Robotics Store

"The world's biggest collection of miniature robots and supplies." This is a great source for servos, tracked vehicles, and robot arms.

Order Desk Mondo-tronics Inc.
524 San Anselmo Ave #107-13
San Anselmo, CA 94960
Tel: 1(800)374-5764
Fax: (415)455-9333
Web: http://www.robotstore.com/

PIC-specific companies

Here are a number of companies that sell PIC products. I have broken up the products/services into specific categories with full company information at the end of the list.

Programmers

- "Introduction to 16C84," programmer, and components: DIY
- General 16C*xx* programmer: DIY
- DonTronics DT.001 programmer: DonTronics
- HTH programmer: High Tech Horizon
- PIC-1a programmer: ITU Technologies
- WARP-3 programmer: ITU Technologies
- WARP-3: Newfound Electronics, DonTronics
- WARP-17: Newfound Electronics
- 5xer programmer: ITU Technologies
- PIC16C*xx* programmer: Parallax Inc.
- PROBYTE programmer: PROBYTE Oy
- ProPic: Tato Computadores

Development tools

- Fuzzy logic development system: Byte Craft Limited
- CCS software prototyping board: CCS
- CESSIM development environment: ITU Technologies
- Linux software for parallel port programmers: Nexus Computing
- Programmer adapters for various PIC package types: Parallax, Inc.
- SimmStick: SiStudio
- PICkle: SuperComputing Surfaces, Inc.

Compilers

- MPC "C" compiler: Byte Craft Limited
- PCB DOS "C" compiler (PIC16C5*xx* and 12000): CCS
- PCM DOS "C" compiler (PIC16C6*xx*, 7*xx*, 8*x*, 92*x*, and 14000): CCS
- PCW Professional Package for Windows (all PIC16C*xx* chips): CCS
- FED PIC Basic compiler: DonTronics

- MEL PIC Basic compiler: DonTronics
- ANSI C compiler for the PIC: HI-TECH Software
- PIC BASIC compiler: ITU Technologies
- PIC assembler: Parallax Inc.
- C compiler: Parallax, Inc.

Emulators

- Emu emulator: Wirz Electronics, DonTronics, Interface Products (Pty) Ltd.

Applications

- PIC-based closed caption decoder: Brouhaha Computer Mercenary Services
- Radio-based data collection system: Cedardell Inc.
- DIX: Digital Intelligent Cross Switch: CherryTronics B.V.
- CAPS: CherryTronics Automatic Power Switch: CherryTronics B.V.
- COPS: CherryTronics OS/2 Power Switch: CherryTronics B.V.
- TELEPHONEGUARD: CherryTronics B.V.
- Disk drive test power supply: Diamond Mountain Engineering, Inc.
- Single and dual dice: DIY
- Unipolar stepper motor controller: DIY
- Servo motor driver: DIY
- PICPlus™ Microcontroller Board: E-Lab Digital Engineering, Inc.
- Alarm and monitoring system: Engenharia Mestra de Sistemas, sociedade limitada
- Unipolar 3-, 6-, or 8-wire stepper motors: Fisher Automation, Inc.
- DC load switching: Fisher Automation, Inc.
- Motor control: JS Controls
- DigiTemp temperature sensor system for DOS and Linux: Nexus Computing
- RS-232 to 4-20 mA converter using a PIC: Oak Valley Development
- BASIC Stamp: Parallax, Inc.
- PROBYTE tools: PROBYTE Oy
- GPS-based timing receiver: Rack and Stack Systems
- Mini mods: Solutions Cubed
- SLI-LCD interface: Wirz Electronics, DonTronics, Interface Products (Pty) Ltd.

Publications

- *Byte Craft* quarterly newsletter: Byte Craft Limited
- PIC datasheets, magazine articles, and application notes in Russian: ORMIX Ltd.
- *Easy PIN'n Beginner's Guide*: DonTronics
- Scott Edward's PIC Source Book/Disk: DonTronics

Consultants

- Aengineering Co.
- AmberDrew Ltd.

- Brouhaha Computer Mercenary Services
- Cinematronics
- CherryTronics B.V.
- E-Lab Digital Engineering, Inc.
- Embedded Research
- Engenharia Mestra de Sistemas, sociedade limitada
- Fast Forward Engineering
- Interface Products (Pty} Ltd.
- Iversoft, Inc.
- Jones Computer Communications
- Nelson Research
- Nexus Computing
- Oak Valley Development
- ORMIX Ltd.
- Pragmatix
- Precision Design Services
- PROBYTE Oy
- Rack and Stack Systems
- Rochester MicroSystems, Inc.
- Solutions Cubed
- SuperComputing Surfaces, Inc.
- Tato Computadores
- Telelink Communications

Microchip representatives

- Nelson Research
- ORMIX Ltd.
- Pipe-Thompson Technologies Inc.
- PROBYTE Oy

The companies

Aengineering Co.

We design noncontact sensors for OEMs and users, especially sensors used for sensing people. Successful commercial designs include: lavatory faucet and toilet no-touch controls, intrusion alarms, aiming and focusing devices, and many others. Ultra-low cost consumer product designs make ingenious use of PIC microcontrollers.

3300 S Fox Spi Rd.
Langley, WA 98260-8010
Tel: (360)730-2058
Fax: (360)730-2058
Web: http://www.whidbey.com/optoinfo
Email: optoeng@whidbey.com

AmberDrew Ltd.

AmberDrew specializes in the development of software and hardware solutions in process monitoring and control using embedded controllers based on PICs and PCs.

Thistle Lodge
Alltewn
Pontardawe Swamsea
South Wales, UK SA8 3A!
Tel: +44 (0)1792 862912
Fax: +44 (0)1792 862912
Email: efoc@cyberstop.net

Byte Craft Limited

Byte Craft produces application code development products for 8-bit embedded systems. Byte Craft is noted for its excellence in design and product performance on diverse architectures and for providing innovative solutions to developers, manufacturers and consultants. Byte Craft's products include assemblers, C compilers, and fuzzy logic preprocessors.

421 King Street
North Waterloo
Ontario, Canada N2J 4E4
Tel: (519)888-6911
Fax: (519)746-6751
Web:http://www.bytecraft.com
Email: info@bytecraft.com

Brouhaha Computer Mercenary Services

Embedded system design consulting. A few PIC project designs available free from Web page.

142 North Milpitas Boulevard
Suite 379
Milpitas, CA 95035
Tel: (408)263-3894
Web: http://www.brouhaha.com/~eric/pic/

Cedardell Inc.

A radio-based data collection/delivery system, using a multi-loop architecture. System can be connected to a PC or a microprocessor-based control system via a RS-232 link. 16-bit Windows support DLL and sample Windows management application provided. Currently used in a security system.

2919 17 St
#210
Longmont, CO 80503
Tel: (303)651-2442
Fax: (303)651-2426
Web: http://www.sni.net/cedardell
Email: edtodd@sni.net

CherryTronics B.V.

DIX: An intelligent power-saving device for computer systems, which diminishes power-consumption of PCs, equipped with Advanced Power Management (APM) with more than 99% in Standby Mode.

CAPS: A device used for remotely powering on and off computer systems (used for the unattended exchange of data/software).

COPS: A device for automatic powering on and off computer systems by telephone without answering the call; information is transferred toll-free.

TELEPHONEGUARD: A device that prevents unauthorized use of one's telephone line by illegal callers (a very hot item in our region of the world).

Haarstraat 25-a Ammerzoden, Holland 5324 AM

P.O. Box 39
Ammerzoden, Holland 5324 ZG

Tel: +31 73 5991098
Fax: +31 73 5994712
Web: http://www.cherrytronics.nl
Email: mdekkers@inter.nl.net   b.janssen@inter.nl.net
infor@cherrytronics.nl

Cinematronics

Consultant specializing in the design and fabrication of electronic devices for the film/video and entertainment industry. Custom design of lighting controllers, motion control devices, and remote control electronics. Cinematronics also designs PIC-based electronic devices for consumer electronics.

344 Dupont St
Suite 304
Toronto, Ontario, Canada M5R 1V9
Tel: (416)927-7679
Fax: (416)927-7679
Email: cinetron@passport.ca

Custom Computer Services, Inc.

CCS has created a line of powerful C compilers that are designed specifically for the Microchip PIC16Cxx devices to make development fast, easy, and efficient. Standard C operators and the special built-in functions are optimized to produce very efficient code for the bit and I/O functions normally required for these microcontrollers.

P.O. Box 53008
Brookfield, WI 53008
Tel: (414)781-2794 x30
Fax: (414)781-3241
Web: http://www.execpc.com/~ccs/picc
Email: ccs@execpc.com

Diamond Mountain Engineering, Inc.

We sell a programmable power supply based on a PIC 16C64. It is designed for disk drive testing with a 5-V and 12-V output rated at 3 Amps. It features programmable linear ramping from 10 microseconds to 10 seconds in three ranges. The voltage range is from 0 V to 6 V and 0 V to 16 V. The fast 16C64 allows positive, negative, and bipolar spikes from 10 microseconds to 10 seconds.

3123 Whipple Road
Union City, CA 94587-1218
Tel: (510)487-9530
Fax: (510)487-9531
Email: mschwabe@ricochet.net

DIY Electronics

Manufacturers of electronic kits using PIC IC. All source code supplied on floppy disk.

Products include:
- Single and dual dice
- Unipolar stepper motor driver
- Servo motor driver
- "Introduction to 16C84," programmer, and all components
- General 16C*xx* programmer

P.O. Box 88458
Sham Shui PO
Hong Kong
Tel: 852-2720 0255
Fax: 852-2725 0610
Web: http://www.hk.super.net/~diykit/
Email: diykit@hk.super.net

DonTronics

PCB kits including Microchip Technology PIC-based micro products, FED and MEL PIC Basic compilers, FED PIC Basic interpreters, CCS C Compilers, Square 1 Electronics EasyPIC'n Beginners Guide, Newfound PIC Programmers DonTronics DT.001 programmers, *Nuts & Volts* magazine, Scott Edward's PIC Source Book/Disk, SiStudio SimmStick, and Wirz Electronics SLI-LCD Interface, and Emu PIC 18-pin emulator.

P.O. Box 595
Tullamarine 3043 Australia
Tel: 613+9338-6286
Fax: 613+9338-2935
Web: http://www.labyrinth.net.au/~donmck
Email: donmck@labyrinth.net.au

E-Lab Digital Engineering, Inc.

Our PICPlus™ Microcontroller Board is designed to enhance the PIC16C57 microcontroller, giving it expanded I/O capabilities as well as a direct-connect LCD

port. Included driver routines greatly enhance development speed. Ideal for prototyping, hobbyists, robotics, etc.

> 1932 Hwy. 20
> Lawton, IA 51030
> Tel: (712)944-5344
> Fax: (712)944-5501
> Web: http://www.netins.net/showcase/elab
> Email: elab@netins.net

Embedded Research

Embedded Research offers engineering services, both hardware and software,
for the embedded controller market. We specialize in PIC-based designs.

> P.O. Box 92492
> Rochester, NY 14692
> Tel: (716)359-3941
> Web: http://www.vivanet.com/~gmdsr
> Email: gmdsr@vivanet.com

Engenharia Mestra de Sistemas, sociedade limitada

Engenharia Mestra makes alarm systems for all types of clients, from correc-
tional facilities to the regular home. We have four types of panels in production, and
we also make custom systems, as well as act in consulting.

> Rua Guaiauna
> 439 Sao Paulo, SP
> Brazil 03631-000
> Tel: 55-11-218.5008
> Fax: 55-11-217.6610
> Email: mestra@u-netsys.com.br

Fast Forward Engineering

Contract design of PIC-based products, specializing in RF systems, high-volume
consumer electronics, automotive electronics, and remote telemetry and control.
Software-only services are available, as well as start-to-finish handling of the entire
design process. Consulting/training services are also available.

Certified by Microchip as one of their Microchip Technology Consultant Program
Members.

> 1984 Casablanca Court
> Vista, CA 92083-5043
> Tel: (619)598-0200
> Fax: (619)598-2950
> Web: http://www.geocities.com/SiliconValley/2499
> Email: fastfwd@ix.netcom.com

Fisher Automation, Inc.

Output interface boards available:

- Control of unipolar 5-, 6-, or 8-wire step motors (up to 2 Amps/coil up to 30 VDC) from only two pins.
- General on/off switching of dc loads. Four loads can be controlled up to 2 Amps each to 60 VDC each.

150-45 12 Rd.
Whitestone, NY 11357
Tel: (718)767-8250
Fax: (718)767-8251
Web: http://www.plasticsnet.com/mbr/fishfam
Email: fishfam@pipeline.com

High Tech Horizon

HTH distributes Parallax PIC and Basic Stamp products in Scandinavia. This is also the home of the List of Stamp Applications (LOSA).

Asbogatan 29 C
Angelholm, Sweden S-262 51
Tel: +46 431 41 00 88
Fax: +46 431 41 00 88
Web: http://www.hth.com
Email: info@hth.com

HI-TECH Software

HI-TECH produces an ANSI-C compiler for PIC series microcontrollers. It includes floating point, long arithmetic, and compact code, complete with assembler and linker

P.O. Box 103
Alderley, QLD, Australia 4051
Tel: +61 7 3354 2411
Fax: +61 7 3354 2422
Web: http://www.htsoft.com/
Email: sales@htsoft.com

Interface Products (Pty) Ltd.

Supplier of microcontroller development tools, and custom electronic design and manufacturing. Remote automation, monitoring, and control across commodity networks. Positioning support for GPS assisted airborne remote sensing; stocklist of Garmin equipment. Business level Internet services for engineering professionals. Distributor for DonTronics, Silicon Studio, and Wirz Electronics microcontroller products.

2nd Floor
Quinor Court
81 Beit Street
New Doorfontein Gauteng, South Africa

P.O. Box 15775
DOORNFONTEIN 2028
South Africa
Tel: +27 (11) 402-7750
Fax: +27 (11) 402-7751
Web: http://www.ip.co.za
Email: info@ip.co.za

ITU Technologies

ITU Technologies is a leading supplier of PIC development tools. We provide the highest quality tools at the most affordable prices! Our PIC-n-GO package is a perfect introductory package for only $45 (kit) or $65 (assembled). This includes our PIC-1a programmer, PIC16F84 microcontroller, and an assembler.

3477 Westport Ct.
Cincinnati, OH 45248
Tel: (513)574-7523
Fax: (513)574-4245
Web: http://www.itutech.com
Email: sales@itutech.com

Iversoft, Inc.

Programming services and product consulting.

40 Samoset Ave.
1st Fl.
Mansfield, MA 02048
Tel: (508)337-9926
Fax: (508)337-9926
Web: http:www.iversoft.com
Email: anick@iversoft.com

Jones Computer Communications

Embedded systems design and programming in assembler and/or C. Strong background in networks, serial communications, and radio.

509 Black Hills Dr.
Claremont, CA 91711
Tel: (909)621-9008
Fax: On Request
Email: lee@frumble.claremont.edu

JS Controls

Civil laboratory force testing equipment, manufacture, service, and calibration. Currently offering PIC16C84-based dc motor speed controller:

- 200V shunt wound motors
- 3-digit LED rpm readout
- Push-button stop/start and speed control
- Nonvolatile memory for speed set-point

8 Nolloth St.
Groeneweide Boksburg
1460 Republic of South Africa
Tel: Johannesburg 893-4154
Fax: Johannesburg 893-4154
Email: jsand@pixie.co.za

Nelson Research

Expert in Microchip PIC microprocessors and embedded systems design. Design and coding of assembly language software for embedded systems. Design of analog systems and subsystems, including low-level instrumentation. Troubleshooting and failure analysis of systems and processes. Creation and presentation of custom training courses and seminars.

130 School Street
P.O. Box 416
Webster, MA 01570-0416
Tel: (508)943-1075
Fax: (508)949-2914
Web: http://www.ultranet.com/~NR
Email: L.Nelson@ieee.org

Newfound Electronics

WARP-3 is a high featured, low-cost development programmer for all PIC16Cxx and derivative devices. WARP-3 is a DOS-based programmer and suitable for use with anything from an XT to a Pentium PC. Please see the Web site for further WARP-3 details and also the low cost, fully featured "production" programmer to follow.

WARP-17 is a low-cost development programmer for the 17C4x devices. All configuration options are supported, and the software allows easy code and configuration changes. The WARP-17 is the ideal programmer to evaluate the increased power of the 17C4x family.

14 Maitland St.
Geelong West Victoria 3218 Australia
Web: http://www.labyrinth.net.au/~newfound
Email: newfound@iaccess.com.au

Nexus Computing

Nexus Computing specializes in the development of small embedded microprocessor products using Microchip PIC and Motorola 8- and 16-bit processors.

Web: http://www.eskimo.com/~nexus
Email: nexus@eskimo.com

Oak Valley Development

Oak Valley Development is a group of consultants that specializes in microprocessor controlled electro-mechanic and analog electronics. We have complete engineering resources to do all phases of a project from prototype to high volume production.

P.O. Box 41473
San Jose, CA 95160
Tel: (408)489-9623
Fax: (408)268-6184
Web: http://www.oakvalley.com
Email: paul@oakvalley.com

ORMIX Ltd.

Low-cost development programmers COMPIC-1 and COMPIC-5X for Microchip MCUs and Serial EEPROMs, connected to the serial port. COMPIC-1 works without an external power supply, COMPIC-5X has two ZIF sockets. Excellent software flexibility—new devices can be added to the text configuration file by end user. Very simple and friendly user interface, built-in HEX editor, serialization of parts possible. Low price.

Kr.Barona
136 Riga, Latvia LV-1012
Tel: (371)-7310660
Fax: (371)-2292823
Web: http://www.ormix.riga.lv/eng/index.htm
Email: avlad@mail.ormix.riga.lv

Parallax, Inc.

The Parallax PIC16Cxx programmer can program many PIC devices and is constantly updated for newer PICs throughout the year. We also offer many adaptors that plug directly into the programmer for use with various package types available in the PIC16Cxx series.

3805 Atherton Rd.
Suite #102
Rocklin, CA 95675
Tel: 1(888)512-1024   (916)624-8333
Fax: (916)624-8003
Web: http://www.parallaxinc.com
Email: info@parallaxinc.com

Pipe-Thompson Technologies Inc.

Pipe Thompson is the Microchip rep for the Toronto Area. Along with providing technical information to Microchip customers, Pipe-Thompson can provide technical guidance and support.

4 Robert Speck Pkwy
Suite 1170
Mississauga, Ontario Canada L4Z 1S1
Tel: (905)281-8281
Fax: (905)281-8550
Email: pipethom@idirect.com

Precision Design Services

Precision Design Services is an engineering consortium specializing in developing new products from idea through to a fully manufacturable, low-cost, high-quality product. Specific areas of expertise are: embedded microcontroller hardware and software design and very high speed receivers (in ECL, 500 MBits/sec). PDS has extensive experience in the design and manufacturing processes of printed circuit boards from simple to double-sided SMT and impedance controlled boards.

26006 View Point Drive
East Capistrano Beach, CA 92624-1224
Tel: (714)489-1064
Fax: (714)489-9299
Email: KASPER@EXO.COM

Pragmatix

Pragmatix offers its customers a full-service solution for their PIC-based embedded systems needs. Our capabilities include product specification, hardware and software development, prototyping, volume production, and systems integration. Also, Pragmatix offers unique parameterizable PIC cores for implementation in ASICs and FPGAs.

Bosuil 2 B2100 Deurne Belgium
Tel: (+32) 3-3247733
Fax: (+32) 3-3247733
Web: http://www.geocities.com/SiliconValley/6766
Email: preynen@innet.be

PROBYTE Oy

Products for industrial use: temperature measuring devices, RF/audio test equipment, PCM-test generators, data loggers. Microprocessor development tools: Protel CAD, Needham eprommers, Tech Tools emulators, Parallax PIC and Stamp tools, microEngineering PicBasic, Microchip tools, and PICs.

Nirvankatu 31
Tampere, Finland 33820
Tel: Int+358-3-2661885
Fax: Int+358-3-2661886
Web: http://www.sci.fi/~pri
Email: pri@sci.fi

Rack and Stack Systems

Engineer (MSEE) with extensive experience in RF, microwave, semiconductors, GPS time, and frequency. Programming in PIC Assembler, HP Rocky Mountain Basic, and National Instruments LabView. All phases of system integration.

3425 Deerwood Drive
Ukiah, CA 95482-7541
Tel: (707)463.2380
Fax: (707)462-4333
Email: brooke@pacific.net

Rochester MicroSystems, Inc.

Rochester MicroSystems, Inc. provides electronic design services to companies and research institutions. We provide electronic system design, electronic circuit design (analog and digital), microprocessor and interface systems, embedded systems and software, and printed circuit board layout design. We have expertise in many application areas.

200 Buell Road
Suite 9
Rochester, NY 14624
Tel: (716)328-5850
Fax: (716)328-1144
Web: http://www.frontiernet.net/~nmi/
Email: rmi@frontiernet.not

Solutions Cubed

Solutions Cubed provides embedded systems design, specializing in Microchip Technology microcontrollers. In addition to full design services and capabilities, Solutions Cubed provides a product line of mini mods. These miniature engineering modules are ideal for the electronic hobbyist and can be easily interfaced to Basic Stamps and their ilk.

3029 Esplanade #F
Chico, CA 95973
Tel: (916)891-8045
Fax: (916)891-1643
Web: http://www.solutions-cubed.com
Email: solcubed@solutions-cubed.com

SuperComputing Surfaces, Inc.

PICkle is an EDA tool for assembly programmers. Automates coding for hardware configuration. Supersedes functional libraries. Expert system with built-in wizards that generate peripheral handling code, math functions, precision delays, interrupts handlers, and more. A true automatic code generator that optimizes register usage. Requires Windows.

8681 North Magnolia Avenue
Suite F Santee, CA 92071
Tel: (619)562-5803
Fax: (619)562-3728
Web: http://pickleware.com
Email: info@pickleware.com

Tato Computadores

ProPic programmer is a new programmer working under Windows and can program almost any PIC.

R Santos Arcadio
37 Sao Paulo/SP Brazil 04707-110
Tel: (011)240-6474
Fax: (011)240-6474
Web: http://www.geocities.com/SiliconValley/Pines/6902/index.html
Email: nogueira@mandic.com.br

Telelink Communications

Our company specializes in radio telemetry/SCADA systems, data logging over radio links, and short haul radio modem systems. A design and consultant service is also provided for specialized projects where required.

Nugget Avenue Bouldercombe
Queensland Australia 4702
Tel: 61 79 340413   61 418 7995551
Fax: 61 79 340413
Web: http://www.networx.com.au/mall/tlink
Email: jack@networx.com.au

Epilogue

Now that you have gotten through the book, you probably think you know everything there is to know about developing applications for the PIC series of microcontrollers.

Well. . . . Not quite.

There's probably at least two more books of material that I haven't covered or have missed. As well, Microchip is planning on releasing at least 30 new part numbers of PICs this year (1997). Chances are there will be hardware features that I haven't even touched on. What I hope I have given you is an understanding of the PIC architecture and enough reference sources that will help you puzzle out how to use these new devices and features.

Looking through the projects, you'll see how my understanding of the PIC has improved over time (the earlier applications aren't done as efficiently as the later ones). Somewhere in the neighborhood of 150 programs were written for this book (the examples, the experiments, and the projects) and as I have written them, my understanding of one PI has dramatically improved. Hopefully, some of this knowledge has been passed along.

I would love to hear from you and hear about your questions, successes, and what you have learned. However, rather than contacting me directly, I would appreciate you e-mailing to the PICLIST (instructions on how to access PICLIST are given in appendix D). This way your information will be available for the use of a much wider audience.

There is one more question that I can answer: What does "PIC" stand for?

The acronym PIC stands for *Peripheral Interface Controller*. With the myriad of different ways peripheral hardware can be connected to these devices and the different ways PICs can be connected to other systems as a peripheral, I think you'll agree that this is a very appropriate name for it.

Good Luck and I look forward to seeing what you come up with!

Index

About the Author

Myke Predko is a New Product Test Engineer at Celestica in Toronto, Ontario, Canada, where he works with new electronic product designers. He has also served as a test engineer, product engineer, and manufacturing manager for some of the world's largest computer manufacturers. Mr. Predko has a patent pending regarding the automated test of PC motherboards. He is a graduate of the University of Waterloo in the field of electrical engineering.

SOFTWARE AND INFORMATION LICENSE

The software and information on this diskette (collectively referred to as the "Product") are the property of The McGraw-Hill Companies, Inc. ("McGraw-Hill") and are protected by both United States copyright law and international copyright treaty provision. You must treat this Product just like a book, except that you may copy it into a computer to be used and you may make archival copies of the Products for the sole purpose of backing up our software and protecting your investment from loss.

By saying "just like a book," McGraw-Hill means, for example, that the Product may be used by any number of people and may be freely moved from one computer location to another, so long as there is no possibility of the Product (or any part of the Product) being used at one location or on one computer while it is being used at another. Just as a book cannot be read by two different people in two different places at the same time, neither can the Product be used by two different people in two different places at the same time (unless, of course, McGraw-Hill's rights are being violated).

McGraw-Hill reserves the right to alter or modify the contents of the Product at any time.

This agreement is effective until terminated. The Agreement will terminate automatically without notice if you fail to comply with any provisions of this Agreement. In the event of termination by reason of your breach, you will destroy or erase all copies of the Product installed on any computer system or made for backup purposes and shall expunge the Product from your data storage facilities.

LIMITED WARRANTY

McGraw-Hill warrants the physical diskette(s) enclosed herein to be free of defects in materials and workmanship for a period of sixty days from the purchase date. If McGraw-Hill receives written notification within the warranty period of defects in materials or workmanship, and such notification is determined by McGraw-Hill to be correct, McGraw-Hill will replace the defective diskette(s). Send request to:

Customer Service
McGraw-Hill
Gahanna Industrial Park
860 Taylor Station Road
Blacklick, OH 43004-9615

The entire and exclusive liability and remedy for breach of this Limited Warranty shall be limited to replacement of defective diskette(s) and shall not include or extend to any claim for or right to cover any other damages, including but not limited to, loss of profit, data, or use of the software, or special, incidental, or consequential damages or other similar claims, even if McGraw-Hill has been specifically advised as to the possibility of such damages. In no event will McGraw-Hill's liability for any damages to you or any other person ever exceed the lower of suggested list price or actual price paid for the license to use the Product, regardless of any form of the claim.

THE MCGRAW-HILL COMPANIES, INC. SPECIFICALLY DISCLAIMS ALL OTHER WARRANTIES, EXPRESS OR IMPLIED, INCLUDING BUT NOT LIMITED TO, ANY IMPLIED WARRANTY OF MERCHANTABILITY OR FITNESS FOR A PARTICULAR PURPOSE. Specifically, McGraw-Hill makes no representation or warranty that the Product is fit for any particular purpose and any implied warranty of merchantability is limited to the sixty day duration of the Limited Warranty covering the physical diskette(s) only (and not the software or information) and is otherwise expressly and specifically disclaimed.

This Limited Warranty gives you specific legal rights; you may have others which may vary from state to state. Some states do not allow the exclusion of incidental or consequential damages, or the limitation on how long an implied warranty lasts, so some of the above may not apply to you.

This Agreement constitutes the entire agreement between the parties relating to use of the Product. The terms of any purchase shall have no effect on the terms of this Agreement. Failure of McGraw-Hill to insist at any time on strict compliance with this Agreement shall not constitute a waiver of any rights under this Agreement. This Agreement shall be construed and governed in accordance with the laws of New York. If any provision of this Agreement is held to be contrary to law, that provision will be enforced to the maximum extent permissible and the remaining provisions will remain in force and effect.